CONTENTS

WITNESSES

HIGH PERFORMANCE COMPUTING AND COMMUNICATIONS

TUESDAY, MAY 10, 1994

HOUSE OF REPRESENTATIVES,
COMMITTEE ON SCIENCE, SPACE, AND TECHNOLOGY,
SUBCOMMITTEE ON SCIENCE,
Washington, D.C.

The subcommittee met, pursuant to call, at 9:35 a.m. in Room 2318, Rayburn House Office Building, Hon. Rick Boucher [chairman of the subcommittee] presiding.

Mr. BOUCHER. The subcommittee will come to order.

This morning the Subcommittee on Science continues its oversight of the high performance computing and communications program. This multiagency R&D initiative was authorized by the High Performance Computing Act of 1991. This is the third oversight hearing on the HPCC program which this subcommittee has conducted during the course of the 103rd Congress.

In April of last year, the subcommittee reviewed the planning and management of the overall program. Last October we reviewed how the program relates to the technology needs of industries, such as manufacturing and financial services. Today we will review the fiscal year 1995 budget request for the program and examine the rationale for the funding priorities that are assigned in the budget request.

The HPCC program encompasses four distinct activities: First, to develop technologies needed for high-speed data networking and to provide network access for the research and education community; secondly, to support the development of advanced software technology for solving important problems in science and engineering often referred to as Grand Challenges; third, to stimulate development of new generations of high performance computing systems; and, fourth, to support basic research and human resources development in computer and computation sciences.

The HPCC program plan for the current fiscal year includes a new component, information infrastructure technology and applications. Through partnerships with industry, educational institutions, and State and local governments, computing and networking technologies will be used to develop and demonstrate applications in areas of broad public interest such as education, health care, and improved access to government information.

This applications component anticipates the new authority for the HPCC program provided in H.R. 1757, the National Information Infrastructure Act of 1993, which was developed by this subcommittee and passed by the House of Representatives last July.

It will be addressed as a part of the upcoming conference with the Senate on the National Competitiveness Act otherwise known as S. 4.

The HPCC program is a major part of the Clinton Administration's National Information Infrastructure Initiative. The President's budget request for fiscal year 1995 includes $1.15 billion for the HPCC program, and that is an increase of $217 million or 23 percent above the fiscal year 1994 appropriations level. More than one-half of that growth for fiscal year 1995 is designated for the new information infrastructure technology and applications component of the program.

In today's hearing we are interested in reviewing the steps that are being taken to assess progress toward meeting the goals of the HPCC program and in reviewing how assessments of progress have been made to date and how they have informed the selection of proposed new priorities for program funding. We are also interested in pursuing the connections between the research undertaken in the HPCC program and the technical barriers for deploying the advanced National Information Infrastructure.

We are pleased to have as our witnesses this morning the Director of the National Coordination Office for the HPCC program and the Senior Program Officers responsible for implementation of that initiative in the four agencies whose combined contributions comprise 85 percent of total program funding. The witnesses have been asked to describe the different agency roles for achieving the program's goals, to review the most significant accomplishments of the program to date, and to explain the strategy which led to the proposed budget increases for fiscal year 1995.

I am pleased to welcome our witnesses this morning. We will look forward to your testimony, and we would ask that our panel of witnesses take their seats at the table at this time.

We are pleased to welcome to the subcommittee Dr. Donald Lindberg, the Director of the HPCC Coordinating Office and chairman of the NSTC HPCC Subcommittee—he is in residence at the National Library of Medicine—Dr. Melvyn Ciment, Acting Assistant Director for Computer and Information Science and the Engineering Directorate of the National Science Foundation; Mr. Lee B. Holcomb, Director of the High Performance Computing and Communications Office for NASA; Dr. David Nelson, Associate Director for Energy Research at the U.S. Department of Energy; and Mr. John C. Toole, Acting Director for the Computing Systems Technology Office at ARPA.

Without objection, your prepared written statements will be made a part of the record, and we would welcome your oral summaries of those statements and would ask that you try to keep your oral summaries to approximately five minutes so that we will have time for a dialogue with you.

During the course of your summaries, if you could address the broad questions that were set forth in the letter of invitation and also in these opening comments, that would be most helpful.

Dr. Lindberg, we will be pleased to begin with you this morning.

STATEMENTS OF DONALD LINDBERG, DIRECTOR, HPCC COORDINATING OFFICE AND CHAIRMAN, NSTC HPCC SUBCOMMITTEE, NATIONAL LIBRARY OF MEDICINE, BETHESDA, MARYLAND; MELVYN CIMENT, ACTING ASSISTANT DIRECTOR, COMPUTER AND INFORMATION SCIENCE AND ENGINEERING DIRECTORATE, NATIONAL SCIENCE FOUNDATION, ARLINGTON, VIRGINIA; LEE B. HOLCOMB, DIRECTOR, HIGH PERFORMANCE COMPUTING AND COMMUNICATIONS OFFICE, NATIONAL AERONAUTICS AND SPACE ADMINISTRATION, WASHINGTON, D.C.; DAVID NELSON, ASSOCIATE DIRECTOR FOR ENERGY RESEARCH, U.S. DEPARTMENT OF ENERGY, WASHINGTON, D.C.; AND JOHN C. TOOLE, ACTING DIRECTOR, COMPUTING SYSTEMS TECHNOLOGY OFFICE, ADVANCED RESEARCH PROJECTS AGENCY, ARLINGTON, VIRGINIA

Dr. LINDBERG. Thank you, Mr. Chairman. It is a pleasure to have the opportunity to speak before the committee.

I remain personally enthusiastic about the importance of the HPCC Program goals, the cooperative and creative spirit with which the ten agencies are working together, and about the steady overtaking of the technical objectives.

Because of this relatively good progress, the agencies greeted with enthusiasm this committee's bill earlier in the year, H.R. 1757. We see as an important part of our job to contribute to the scientific and technical platform on which the NII will be built, likewise to the tools and testbeds for the key initial applications on the NII. Such development and testing are essential for perfecting the high performance computer and communications systems.

With this in mind, a new program component, IITA, Information Infrastructure Technology and Applications, has been added to our HPCC overall schema, and support for IITA has been included in each agency's part of the fiscal year 1995 budget request.

On the topic of meetings and advice, we have made substantial efforts this year to increase the amount of information about the program activities that was provided to public and interested non-government experts. This has included the following. All of our reports and all of our solicitations for grants and contracts and announcements have been put on line on a Mosaic server and is available to the public, and it is very heavily used. Special purpose scientific meetings have been held with industry and universities. Examples include the Grand Challenge teams meeting in Pittsburgh, the Pasadena software workshops, the Petaflops meeting in California, and the ARPA symposium here in Virginia. In addition, the NCO has made 80 presentations in the past year to interested associations and groups including, on request, of course, congressional committees and staff briefings. The implementation plan has been made fully public with 125 pages of detailed budget on the 88 major HPCC program elements, and early copies have been provided to the committee. The last annual report has been heavily requested, some 40,000 copies have been asked for and distributed. So all these efforts have put forward information on the program.

Now the counterpart is, of course, for us to listen to comments and suggestions from non-Federal interested persons and groups, and I will mention three special meetings that we arranged in order to get this input. Firstly, U.S. manufacturers of high performance computers; secondly, U.S. telecommunications companies; and, thirdly, U.S. commercial software developers. We have also now scheduled meetings with non-Federal high performance computer center directors and plan to schedule a meeting with academic computer science and information science professionals.

You may well wonder what have we learned from all of this, and I would like to cite two particular things. Firstly, we have found that the HPCC provides a remarkably good mechanism for cooperative scientific research and development amongst Federal agencies. A number of successful management mechanisms have been deployed. I will mention just three: Joint interagency planning and funding of research and infrastructure grants; secondly, the gigabit testbed program, a very interesting program; and, thirdly, consortia of a variety of sorts between Government, industry, and universities.

A second thing we have learned, industry strongly supports the HPCC Program, but it has its own worries, and I will mention three sets of worries. The computer—and we learned this from our own meetings with the—in listening mode with the industry—the computer manufacturers assure us that whereas they couldn't say it three years ago, they now can say that they know they will be able to achieve teraflop computing machines and they are committed to do this. They ask that the Federal Government not deviate from the goal which it has set.

What they are worried about is applications software: Will there be programs, more particularly third party provided programs, once this gear is ready to be sold? The telecommunications industry is cooperating, and coventuring really, with HPCC in a remarkable fashion in these gigabit testbed sites. They are investing far more money than the Federal Government is, and apparently quite happily.

We had a statement in the meeting we had that I referred to from Bell Atlantic in which they stated that they are already marketing products which derive directly from the gigabit testbed participation, in spite of fact that gigabit transmitivity is still not a good profit making business; the Government is still the first big customer in that.

The software industry is also very pleased with the program, and they particularly praised the high performance Fortran development which was funded by HPCC at Rice along with some nine companies. More needs to be done, and in our initial set of meetings they couldn't reach consensus over what that should be, but among the things considered was support for parallel versions of yet other languages, and we all agreed that that was a high priority question and that would be followed up vigorously in the course of the year. So we will attend to that.

I would be happy to answer any further questions that you might have, Mr. Chairman.

[The prepared statement of Dr. Lindberg follows:]

HPCC

Statement of
Dr. Donald A. B. Lindberg
Director
National Coordination Office for
High Performance Computing and Communications

on
The High Performance Computing and Communications Program

Before the

House Committee on Science, Space, and Technology

Subcommittee on Science

May 10, 1994

National Coordination Office for High Performance Computing and Communications
Office of Science and Technology Policy
Executive Office of the President

8600 Rockville Pike • 38A/B1N30 • Bethesda • Maryland • 20894
Phone (301) 402-4100 • FAX (301) 402-4080 • E-mail nco@hpcc.gov

Mr. Chairman and Members of the Subcommittee:

Thank you for inviting me to speak before the Subcommittee about the High Performance Computing and Communications Program. Many of you on the Subcommittee — Representatives Boucher, Brown, and Valentine — were early champions of the HPCC Program, which was established by the High Performance Computing Act of 1991 (P.L. 102-194). The original goals of the Program are to extend U.S. leadership in high performance computing and networking technologies; to disseminate the technologies both to speed the pace of innovation and to serve the national economy, national security, education and the global environment; and to spur gains in U.S. productivity and industrial competitiveness. Strategic objectives include the development of teraops computing systems capable of performing trillions of operations per second and the demonstration of gigabit networks, high speed networks capable of moving billions of bits per second.

This time last year, I spoke before the Subcommittee about H.R. 1757, the National Information Infrastructure Act of 1993, introduced by Chairman Boucher. That bill, which passed the House last summer, embodies the second phase of the HPCC Program.

H.R. 1757 and other legislation introduced into the Congress last year called for the establishment of a National Information Infrastructure (NII or "Information Superhighway.") Applications are identified in the areas of health care, education and lifelong learning, digital libraries, advanced manufacturing, improved access to government information, and other areas.

The HPCC Program is providing the technological foundation upon which the NII is being built. When I spoke before you last year, I reported that the Administration had just added a new Program component — Information Infrastructure Technology and Applications. The addition of this new component brings with it new objectives of developing and demonstrating NII technology, demonstrating National Challenge applications such as those in H.R. 1757 that I just mentioned, and training the Nation to use the NII.

Nine agencies are formal participants in the HPCC Program: the Advanced Research Projects Agency (ARPA), National Science Foundation (NSF), Department of Energy (DOE), National Aeronautics and Space Administration (NASA), National Institutes of Health (NIH), National Institute of Standards and Technology (NIST), National Security Agency (NSA), National Oceanic and Atmospheric Administration (NOAA), and Environmental Protection Agency (EPA). The Department of Education (ED) has allocated resources to HPCC related activities, but this funding is not included in the HPCC Program's budget. These ten agencies comprise the High Performance Computing, Communications, and Information Technology (HPCCIT) Subcommittee.

A number of agencies have observer status, meaning that they attend HPCCIT meetings and plan to work with the HPCC agencies on selected projects. Current observers include the Department of Agriculture, the Agency for Health Care Policy and Research, Centers for Disease Control and Prevention, Food and Drug Administration, Indian Health Service, National Communications Systems, National Emergency Preparedness, and the Department of Veteran Affairs. This number continues to grow.

The National Coordination Office reports to the White House Office of

Science and Technology Policy. As Director of the NCO, I serve as a Special Assistant to the President's Science Advisor, Dr. John Gibbons. HPCCIT now reports to the Committee on Information and Communications of the National Science and Technology Council. The NSTC is trying to do for the entire government what we have already done in the area of high performance computing and communications -- better coordinate and prioritize federal R&D programs.

You have asked me to review the steps being taken to assess progress toward meeting the HPCC Program goals and how assessments of progress made to date have informed the selection of new priorities for Program funding.

The review process consists of ongoing monitoring of progress toward accomplishing Program goals. Progress is reported up from R&D staff through the management chain within individual agencies and joint project management teams.

The HPCCIT Subcommittee meets once a month to coordinate agency HPCC programs, develop interagency programs, and review the plans and activities of participating agencies. Meetings are attended by additional representatives from other Federal agencies and the Administration who are interested or involved in high performance computing and communications. Key policy decisions and plans for HPCCIT meetings are made by an Executive Committee comprised of a subset of HPCCIT members.

Groups that report to HPCCIT coordinate Program efforts in areas such as applications software, networking, and education and training. The Information Infrastructure Technology and Applications Task Group was established in 1993. A February report by the group describes the HPCC Program's new IITA activities. The Task Group has also developed coordinated agency plans for R&D in NII technology and National Challenge applications.

The Networking Infrastructure and Digital Communications Group coordinates network integration activities and works closely with the Federal Networking Council (FNC), whose membership extends beyond HPCCIT. The FNC establishes strategy and policy direction, provides further coordination, and addresses technical, operational, and management issues of the Interagency Internet. Its Federal Networking Council Advisory Committee represents a broad spectrum of network users and providers.

The Science and Engineering Computing Group coordinates activities related to Grand Challenge applications, software tools needed for applications development, and software development at high performance computing center.

The Education for High Performance Computing Group coordinates HPCC education and training activities and provides liaison with other government education-related efforts.

HPCC is a research and development program. Each new set of technologies forms the foundation for the next. As high performance computing hardware is built, the systems software to make it usable is developed, then the applications software. As advanced networking technologies are developed and deployed, software developers at remote locations access this hardware by using systems software to develop applications. Users access one or more high performance systems across this computation-intensive applications software.

This work doesn't proceed absolutely sequentially — systems software development may sometimes begin before hardware is available, for example. Applications programs cannot be executed on multiple systems until the systems and networks are in place. We need progress on all fronts because all these areas are so closely linked.

Mr. Chairman, you once asked me what I consider the most significant accomplishments of the HPCC Program. I believe the most important accomplishment of the HPCC Program is that agencies are collaborating effectively on extremely important scientific projects. Agencies don't always consider working together. And when they do, they often lack the legal mechanisms and management mechanisms to do so.

The HPCC crosscut provides a Congressionally authorized means for our agencies to work together. This includes a variety of management mechanisms. Joint funding of projects, consortia, and cross agency reviews of grants and contracts, and the gigabit testbeds are a few of the mechanisms we have tried.

Joint funding of projects allows for their careful consideration, evaluation, management and followup. Examples of successful joint projects in the HPCC Program include the federal portion of the Internet. In fact, without the HPCC Program, we could have as many as nine different government networks, none of which was interoperable with the other. Joint multi-agency support of the centers and network have led to the next logical level of abstraction: namely, that of a countrywide Metacenter.

The Program supports seven "gigabit testbeds" where R&D in advanced networking technologies is conducted. These technologies include ATM/SONET (Asynchronous Transfer Mode/Synchronous Optical Network), interfacing ATM to HiPPI (High Performance Parallel Interface) and HiPPI switches, and all-optical networking. A key feature of these testbeds is that each addresses an application that requires gigabit speed networks. Some of these have already demonstrated gigabit speeds in an R&D setting.

A variety of consortia formally bring together agencies, federal laboratories, universities, and industry. These include joint efforts between agencies, national laboratories and academia such as the Consortium on Advanced Modeling of Regional Air Quality (CAMRAQ), which includes six federal agencies and a number of state and local governments.

The crosscut mechanism doesn't compel agencies to conduct these activities, but it provides a number of good mechanisms for productive collaboration. For example, agencies also assist each other in cross agency reviews of major grants and contracts where these are of concern to the overall HPCC Program.

As you know, there are significant differences in the way agencies work. Individual agencies support research efforts through various mechanisms. For example, all research supported by the National Science Foundation is in the form of grants and other extramural funding mechanisms. At the Department of Energy, the majority of research funds (approximately 60 percent) go to national laboratories, and about 40 percent to academic grantees. At the National Institutes of Health, about 11 percent of the budget goes for intramural research at the agency — the rest goes to outside investigators.

Despite the different funding approaches, research efforts by these agencies can be and are complementary. They don't have to change the way they support their research to be able to work together because of the flexible management mechanisms employed by the Program. Cosponsorship and cofunding by agencies of important R&D efforts continues. For example, in late 1993, NSF, ARPA and

NASA issued a joint request for proposals for research on digital libraries that will provide some $7 million in combined funds for research, prototype, and testbed activities.

The HPCC Program serves in many ways as a model of a "virtual agency" in which government agencies work together toward common objectives. The crosscut mechanism enables interagency coordination and helps ensure that funds are wisely spent.

The National Coordination Office celebrated its first birthday in September. The office coordinates HPCC programs across Federal agencies through providing traditional management tools and services to participating agencies. The NCO hosts regular interagency meetings as well as internal and external briefings. The office also acts as a liaison to Congress, state and local governments, foreign governments, industry, universities, and the public.

We believe our overall Program goals and objectives relate well to the needs of the computer and telecommunications industries as well as the industries that will be users of high performance computing and networking. We actively seek input from industry and other sectors as to their needs and advice.

In the past year, the HPCCIT Subcommittee has initiated a series of open meetings with key private sector experts and affected groups in order to obtain more public input into our decisions concerning program direction.

These meetings have provided invaluable advice and insight. For example, when we hosted our first such meeting last November with leaders of the computer hardware industry, we learned that their major concern was not whether they could reach the Program goal of achieving teraops performance. They are certain they will and urged that the Federal government not alter its goals. What they worry about is whether applications software will be developed that can take full advantage of these remarkable advances in computational power that can help address so many of the Nation's problems.

Since then, we have already launched new efforts in software development. Because NASA leads the HPCC activities in software development, I will defer to Lee Holcomb, Director of NASA's HPCC program, for a discussion of those initiatives.

Because parallel computing accounts for a very small percentage of the more than $100 billion per year computing market, incentives or special projects may be needed to assure the development of commercial applications.

At our February meeting, we invited representatives from software developers to give their views about the HPCC Program. As to be expected, their concerns were more diverse. They did express enthusiasm for the recent inclusion of the IITA component, particularly National Challenge applications that are closer to market needs. The software representatives also called for software development tools that can be shared, as well as the development of standards to facilitate the exchange of data and software between high performance platforms.

We have also met with representatives from the telecommunications industry, who recommended the government continue in its role as a catalyst for new technologies through existing projects such as the gigabit testbeds. They want government to work with industry to develop standards for interoperability. An open architecture and interoperability were cited as critical elements in the development of a successful NII, which will be able to accommodate wired, wireless, satellite, and fiber optic communications.

The Administration is committed to creating an advisory committee to the Committee on Information and Communications of the National Science and Technology Council that will provide advice on priorities for HPCC. The White House is in the final stages of selecting members of that advisory committee.

Products whose development the HPCC Program supported are now being used by American industry. For example, the Ford Motor Company is now "crash-testing" vehicles via computer simulation on high performance systems, significantly reducing the time for introduction of new models. The Oracle Corporation reports it has installed its data base management software on virtually every high performance system. Already, technologies developed in HPCC supported gigabit testbeds are being made commercially available.

We also obtain feedback through regular workshops, conferences, and public meetings sponsored by participating agencies and the HPCCIT. In 1993, we participated in or hosted more than 80 such events.

These include a Petaops workshop held in February of this year. Now that systems capable of sustained teraflops performance are on the horizon, this workshop looked to the future to address the technological developments needed for three more orders of magnitude faster speed and the applications that will require such speed.

The first Pasadena workshop, on "Systems Software and Tools for High Performance Computing Environments," was held in April 1992. A second Pasadena workshop is scheduled for this coming September. This will focus on how best to address applications software issues, such as how to encourage the development of commercial applications software for high performance systems. These meetings and interactions help us identify Program strengths and weaknesses, and assess progress toward our goals.

I'd like to discuss progress made in the HPCC Program. When it comes to progress in meeting our technical objectives of computing and network performance, we are ahead of schedule. Computers are bigger and faster. Parallel architecture works, and the software is sufficient for initial proof-of-concept experiments.

We are only an order of magnitude away from reaching the original Program goal of achieving teraops performance. Already, HPCC supported architectures are reaching performance levels of 50, 100 gigaops and even 150 gigaops.

Like any other program, of course, we can go only as fast as our weakest or slowest link. This is especially evident in our National Research and Education Network (NREN) component. Internet now extends across the country and around much of the world. By the beginning of FY 1994, almost 2 million computers were accessible over Internet; more than 15,000 regional, state, and local U.S. networks and more than 6,000 foreign networks in approximately 100 countries were part of the Internet. Nearly 1,000 4-year colleges and universities, 100 community colleges, 1,000 U.S. high schools, and 300 academic libraries in the U.S. were connected. Internet continues to grow at a phenomenal rate.

But ironically, our own successes will challenge this. One of our High Performance Computing Research Centers, the National Center for Supercomputing Applications, has developed Mosaic, what The New York Times called the first "killer ap," that is, an extremely successful application in the National Information

Infrastructure. Thousands of servers around the world can be accessed by using Mosaic, which provides hypertext, graphics, audio and video capabilities. This has proved to be an extremely popular application, but it also challenges our network capacity.

New network architectures will relieve pressures by providing better connections. But we will have problems if we don't get the procurements for critical extensions to the NREN through.

Strategic planning is an integral part of the HPCC Program. We prepare an annual report to Congress known as the Blue Book as required by the High Performance Computing Act. We also prepare an annual "Buy American" report for Congress that identifies foreign procurements.

One of the most useful tools we have for assessing progress and selecting priorities for program funding is our Implementation Plan, which was just completed last month.

We've had implementation plans in the past, but never with this level of detail. This year, for the first time, we have made the report public. You and your staff were provided early copies of the plan.

The process of developing this Implementation Plan has been extremely useful. We began by holding a meeting last September at which representatives from each agency made detailed presentations about their individual HPCC program activities. These presentations formed the basis of the Implementation Plan.

The plan identifies 88 project areas, and hundreds of milestones. It identifies interdependencies between agencies, and the activities they fund. For example, all

HPCC agencies depend on advances in networking technologies, so funding network R&D is critical to Program success.

The FY 1994 HPCC Program budget was $937.9 million. The President's budget requests $1.154 billion for FY 1995, a 23 percent increase. FY 1995 represents the first year of funding for IITA activities by all HPCC agencies. The budget for IITA in FY 1994 was $156.0 million, while the request for FY 1995 is $281.6 million. This represents an 80 percent increase. In the last year, the HPCC Program has focused very heavily on adding the IITA component to support the President's NII initiative.

Mr. Chairman, you asked me about the reason for the large variations among proposed budget increases across the participating agencies. The budget request reflects the maturation of the Program and its additional responsibilities to accelerate the development of technologies for the NII. The larger percent increases are for smaller agencies -- the four largest percent increases are for NIST, NOAA, EPA, and NIH.

NIST's increase is primarily for IITA activities, and specifically for advanced manufacturing, a provision in H.R. 820, "The National Competitiveness Act of 1993" introduced by Rep. Valentine. Manufacturing has long been an area of substantial activity at NIST. The new activities will primarily be in systems integration for manufacturing applications.

The NOAA request has two new initiatives -- one to support overall improvements and growth in the agency's Internet connectivity and the other to add a high performance system at its Geophysical Fluid Dynamics Laboratory (GFDL) for research in climate change and weather forecasting. Internet connectivity supports the dissemination of government information, another activity singled out in H.R. 1757. The EPA increase is primarily for acquiring a high performance system for environmental modeling research.

The fourth largest percent increase is for NIH. Most of the increase is for the Medical Connections Program (to increase Internet connectivity to academic medical centers and health care organizations) and health care applications of the National Information Infrastructure. These include testbed networks, and a number of important areas that together are often referred to as telemedicine.

Much of the increases for ARPA, NSF, and NASA support new NII-related responsibilities. These include R&D in NII technologies and digital libraries. The smallest differences are for DOE and NSA.

The variation in proposed HPCC agency budgets is driven by a combination of targeted new initiatives balanced by internal agency priorities. During the summer, the HPCCIT held meetings during which we reviewed individual agency activities and budget submissions and discussed priorities in terms of the overall Program and individual agency missions. We then made funding recommendations, which helped form the basis for the Administration's proposed FY 1995 budget.

HPCC activities to be initiated under the proposed FY 1995 budget will advance the technical goals for the program in a number of ways. Planned communications initiatives will expand the scope and performance levels of networks. Upgrading of the NSF very high speed backbone as well as agency networks in DOE, NASA and NIH will allow sustained responsiveness of the Internet as the number of users and their demands continue to grow. This represents an intermediate step toward the deployment of the high speed technologies needed to support teraops computing and the large number of sophisticated users of the NII. Internet connectivity will be expanded to include 30 NOAA data centers and the K-12 program will increase by more than 50 percent the number of schools connected to the Internet to 1500.

Our high performance computing initiatives will continue to develop the underlying technologies of scalable high performance computers to push toward the goal of enabling teraops computing in 1996. A shift has begun to take place, however, toward developing the system software and software environments that will allow more effective and efficient utilization of high performance computing resources. The broad range of applications supported and the shift to include supporting software will ensure that high performance computing systems can be broadly useful to the American economy.

We also plan to acquire more high performance computing resources at selected sites. These include two 150 gigaops systems in DOE, the acquisition of a high performance computing system for NOAA's Geophysical Fluid Dynamics Laboratory at Princeton, NJ, and the acquisition of a scalable parallel system by EPA to support more complex multipollution and cross-media impact and control assessments.

Support for Grand Challenges continues to grow with improvements made in underlying software and growth in high performance systems that have allowed use of longer time period models because of faster execution times. Models have been improved through higher resolution, giving useful results where it was not even possible to execute some models before. These improvements have allowed more realistic physics to be studied, in areas such as weather forecasting. More powerful high performance computers have also allowed increasingly elaborate multidisciplinary models such as climate studies that combine atmospheric and oceanic modeling and pollution models that incorporate air and water pollution models. Modeling of nutrient loading in the Chesapeake Bay and PCB transport in the Great Lakes are being targeted this year.

Grand Challenge applications support will also improve through benchmarks now being performed that will help determine where and even why a model will run faster.

As Grand Challenge applications are demonstrated, scalable parallel architectures, tools and environments are improved upon, and our understanding grows, spinoff commercial applications are beginning to be moved to these machines. It is the success of these applications that are necessary to the commercial viability of high performance computing.

Mr. Chairman, I thank you for this opportunity to talk about the HPCC Program. I would be glad to answer any questions you may have.

Mr. BOUCHER. Thank you, Dr. Lindberg. We will have questions after the other witnesses have provided their statements.

Dr. Ciment, we will be pleased to hear from you.

Dr. CIMENT. Thank you, Mr. Chairman, for this opportunity to testify here on behalf of the Foundation's budget request for high Performance Computing and Communications.

The Foundation's HPCC program fosters the full range of advanced computing, communications, and information technologies and is strongly linked to fundamental research which is a prerequisite to the development of future systems and an advanced National Information Infrastructure, the NII.

Let me say at the outset that NSF's accomplishments and plans have benefitted immeasurably from the interagency HPCC coordination process. In this context, NSF coordinates the broad deployment of the interagency Internet, supports basic research and infrastructure and education and training, and is helping to lead in developing a concerted effort in the new component that you mentioned.

To briefly convey a sense of NSF's role in meeting the goals of the program, I will focus on several illustrative examples. NSFNET net is a major component of the interagency NREN program. It was established in 1986 primarily to allow for exchange of information and access to remote resources within the research and education community. Since the backbone network was launched, its traffic has doubled each year, and its transmission capacity has increased more than 30-fold to 45 megabits, 45 million bits per second. NSFNET has grown to become a backbone computer network that links more than 1,200 universities and other research and education institutions to the Internet and through the international connections program links more than 70 countries. The impact of NSFNET on the computer and communications industry has been profound.

Prior to 1986 few networks existed within the research and education community. The introduction of NSFNET stimulated the rapid deployment of regional networks throughout all 50 States and also a need for higher speed connectivity. This in turn stimulated development of whole new industries in the area of data communications networks. The overall result has been that the U.S. is now the leader in both the technology and the deployment here.

NSF has recently begun an orderly transition from the current NSFNET backbone system to a new architecture that will encourage advances in both cutting edge experimental efforts and in broadening the participation of emerging public commodity networks.

While the research and education communities take advantage of the excellent commodity services available from commercial suppliers, NSF will again emphasize advanced networking by supporting applications on the very new high speed Backbone Network Services, the VBNS, in fiscal year 1995.

Turning to the NSF supercomputer centers, they were originally established to provide access to researchers in universities. Today they have become recognized leaders in the national HPCC research and development effort influencing a broad range of scientific disciplines far beyond the academic research community.

The centers have pioneered partnerships with the private sector to apply the technologies and practices to specific design and manufacturing problems, and they have alliances with over 88 companies in many sectors of U.S. industry.

Two years ago, the four NSF centers formed a cooperative effort, which they call the Metacenter, to facilitate effective collaboration and interoperability. In fiscal year 1994 NSF initiated support for Metacenter Regional alliances to encourage inclusion of new providers and users. In fiscal year 1995 NSF expects to double the number of these regional alliances from six to 12.

The NSF HPCC program incorporates the broad participation of more than 70 related disciplines including mathematics, physics, biosciences, engineering, geosciences, and economics. While HPCC is primarily an enabling tool for these other disciplines, it often leads to identifying new approaches and models and problems of fundamental importance to science and engineering. These so-called Grand Challenge applications are problems that are beyond the scope of a single research, a single scientific discipline, and sometimes a single super computer. Just try to predict the weather, design drugs, or analyze the environment, and you come up against massive computation requirements. The NSF Grand Challenges activity encourages an integrated multidisciplinary approach to solutions.

NSF together with ARPA has launched 16 Grand Challenge projects including studies of geophysical fluid dynamics, ground motion in earthquakes, the formation of the universe, biomolecular modeling, and so on. Between four and six additional Grand Challenge application groups will be initiated in fiscal year 1995.

Turning now to the new component, the Information Infrastructure Technology and Applications that has been in response to the White House-led initiative on the National Information Infrastructure, NSF has cochaired a multi-agency report to plan a program that will build on the advances in the computationally intensive parts of the original HPCC program to link up with the information intensive parts in the requirements of the National Challenges problems such as health care, manufacturing, and lifelong education.

In 1994 NSF initiated a program in digital libraries research jointly supported by two of my partners at the table, ARPA and NASA, and National Challenge programs in civil infrastructure, manufacturing and health care systems delivery.

In summary, the NSF HPCC program greatly enhances the conduct of science and engineering by enabling the solution of more complex models and processes by fostering more frequent collaboration across the disciplines, and by providing shared access to large collections of data and information, and moreover by encouraging multi-agency and multisector collaborations.

Mr. Chairman, I believe that the NSF accomplishments and plans that I have sketched suggest how close interagency partnerships and academic-industrial partnerships have contributed to making our program more successful. I would be happy to expand on these remarks here or for the record.

[The prepared statement of Dr. Ciment follows:]

TESTIMONY FOR THE HOUSE SCIENCE AND TECHNOLOGY COMMITTEE

DR. MELVYN CIMENT
ACTING ASSISTANT DIRECTOR
COMPUTER AND INFORMATION SCIENCE AND ENGINEERING
NATIONAL SCIENCE FOUNDATION
MAY 10, 1994

Thank you for the opportunity to testify here today on the Foundation's FY 1995 budget request for the High Performance Computing and Communications program. I am pleased to appear here with my colleagues from the National Coordination Office and other agencies to describe and explain this closely coordinated program.

The National Science Foundation's High-Performance Computing and Communications (HPCC) program fosters the full range of advanced computing communications and information technologies. It includes state-of-the-art production supercomputer systems, special-purpose and experimental systems, high-speed networks that transmit billions of data bits per second, large-scale parallel processing systems to solve immensely complex problems, advanced algorithms and software environments, and visualization systems to depict phenomena ranging from the molecular to the meteorological.

Most importantly for the future, NSF HPCC programs are linked to fundamental research on computer and information science and engineering, the prerequisite to the development of future high-performance computing, communications and information systems.

In short, NSF's HPCC program benefits a broad realm of science, engineering, industry and education as it extends U.S. technological leadership and accelerates the wide dissemination and application of high performance computing and communications and information intensive technologies. In support of the federal HPCC program goals, it seeks to speed the pace of innovation in science and engineering, to improve the ability to investigate fundamental problems in science, and to improve U.S. productivity and industrial competitiveness

Achievements and goals such as these are possible because the overall federal HPCC Program is a closely coordinated partnership between ten agencies, including NSF, working together with national centers and laboratories, technology vendors, industry, and universities.

Each agency has responsibilities that cut across the five components of the overall HPCC program and are coordinated with those of relevant other agencies. NSF's key roles are to coordinate the broad deployment of the Interagency Internet, to support basic research and infrastructure activities, and to foster balance in research participation, education and training. It has a significant role in the new Information Technology Infrastructure and Applications (IITA) component of the HPCC Program.

NSF has six basic inter-related, integrated HPCC program objectives which build on the historic strengths of the agency and take advantage of its strong base activities. The first two focus on advancing the computing infrastructure for science and engineering researchers by supporting national access to high speed networks and to supercomputers. The others derive from the NSF role in the discovery and transfer of pertinent knowledge, and the education and training of people who can apply the new capabilities to a broad array of problems and issues -- academic, industrial, and economic. All objectives, sometimes directly, frequently indirectly, address needs of the computer and information processing industries and the manifold users of these capabilities.

- The first objective is to develop, provide and support advanced research and education networking services and capabilities for connecting researchers, educators and students in universities, high schools, research laboratories, libraries, and businesses. This supports the entire HPCC enterprise effort of building a national information infrastructure.

- NSF's second objective is to provide access to state-of-the-art high performance computing environments and to incorporate new generations of scalable parallel high performance computers and software technologies into important application areas. These systems will have the potential to perform one trillion computer calculations per second to help in the solution of major scientific and engineering problems. This effort helps vendors see the strengths and weaknesses of their products and users see the types of problems those products can effectively address.

- The third objective is to generate fundamental knowledge that can lay the foundation for future advances in high performance computing and communications. Capabilities resulting from new knowledge uncovered by basic research have made dramatic impacts on the offerings of the computer industry, on the power and extent of use of computers, and on the range of applications of the technologies.

- The fourth objective focuses not on processors nor on packet-switching, but on a most important part of the HPCC equation: people. NSF is committed to creating a cadre of scientists, engineers, technical personnel, and educated citizenry who are prepared to take advantage of these new capabilities.

- The fifth objective is to enhance HPCC innovation, technology transfer and U.S. productivity and industrial competitiveness. By encouraging academic-industrial partnerships, NSF Centers' affiliates programs contribute to addressing the needs of U.S. industries. For example, last year at the Center for Research on Parallel Computation at Rice University, at least seven companies participated in the development of a standardized specification for a High-Performance Fortran language, and at least 12 others have announced they will offer products based on new standards. Fortran has since the 1950s been the most widely used scientific programming language. High-Performance Fortran, now past the infancy of Version 1.0, could help break down one of the most formidable obstacles to progress in the HPC industry by providing a common, standardized way to express data-parallel programs on a variety of machines.

- The last, but certainly not least, of NSF's objectives is to make advanced computing and communications information infrastructure available to -- and usable by -- a larger segment of the society. By demonstrating solutions to problems with broad societal impact, these efforts may have the most impact on industrial markets because of the large number of people affected

NSF HPCC PROGRAM ORGANIZATION

Let me say a few words on how NSF organizes and manages its HPCC activities. Because the HPCC program crosses disciplinary lines and often requires close collaboration, within NSF an HPCC Coordination Committee facilitates effective operation of the multi-directorate program at a program level, where the day-to-day decisions are made. This group provides advice on priorities, budget levels, and internal management, including the formation of special joint working groups of the discipline-relevant directorates to carry out the detailed research management. The lead responsibility for NSF's HPCC program rests with the Assistant Director for the Directorate of Computer and Information Science and Engineering (CISE).

NSF STRATEGY: ACTIVITIES AND EXAMPLES OF THEIR IMPACT

The overall Federal HPCC Program is often described in terms of its five broad components: High-Performance Computing Systems, the National Research and Education Network, Advanced Software Technology and Algorithms, Information Infrastructure Technology and Applications, and Basic Research and Human Resources. To convey a clearer understanding of the NSF focus in meeting the goals of its HPCC Program, this presentation, however, focuses on several major activities. These are the NSFNET, supercomputer centers, research centers and infrastructure, grand challenge applications, disciplinary HPCC research, education and training.

I will describe each activity briefly and provide examples of its impact on, for example, research and education, design and engineering, and the industries that develop hardware and software for high performance computers and communications. In addition, I will indicate emphases for the FY 1995 budget request.

NSFNET

The NSFNET program is a major component of the Interagency Internet, which in turn is the network infrastructure component of the National Research and Education Network (NREN) program. Let me say at the outset that our accomplishments and plans for the future have been and are critically dependent on and close cooperation of the many agencies involved.

NSFNET's most visible component is the backbone service it provides to link mid-level networks which, in turn, connect commercial and university networks. As the world's fastest computer network available to the general research and education community, this backbone service serves as a major component of the Internet. It provides access to NSF's supercomputing centers and about two dozen regional networks that interconnect thousands of government, academic and health institutions, and private research and development organizations.

NSFNET was established in 1986 chiefly to allow exchange of information and access to remote resources within the research and education community. Since the backbone network was launched, its traffic has doubled each year and its transmission capacity has increased more than 30-fold to 45 megabits--45 million bits--per second. Through its international connections program, NSFNET reaches a worldwide complex of networks linking more than 70 countries.

In the United States, the NSFNET program has had a dramatic impact on the way science is conducted. Prior to 1985 collaborations between colleagues in different institutions were difficult, time-consuming and costly. That changed as NSFNET backbone network services and the associated regional networks extended connectivity throughout institutions in the U.S., and scientists very quickly discovered the ease with which they could communicate with their colleagues, exchange data and results, and seek out new information. As early as 1990, Internet connectivity had become an essential tool for the conduct of scientific research.

Now, in 1994, as editors try to squeeze "information superhighway" into headlines, the NSFNET program's effects are nearly ubiquitous. Basic network information is readily available. Discovery and retrieval tools are routinely used to search out and retrieve documents and information from throughout the Internet world, including those from government agencies. Parenthetically, one of the most popular tools to emerge in this context is the MOSAIC package developed at the National Center for Supercomputing Applications, an NSF supercomputer center. Journals and newsletters are proliferating on the Internet. Multi-institute and multi-disciplinary and multi-location research is becoming more prevalent. Video-conferencing is being implemented. Speeds in excess of 100 megabits per second between supercomputers and users will soon be available. Libraries are making ever increasing amounts of information available electronically.

With enormous implications for the future, access to NSFNET and the Internet -- much of it state-initiated -- for the precollege education community is increasing. Some of the promise is being explored through NSF supported projects. For example, Project DO-IT (Disabilities Opportunities Internetworking Technology) at the University of Washington provides an Internet connection to 20 disabled high school students interested in science and engineering, some of whom are in remote rural areas. These students are linked with a practicing engineer or scientist who shares both their interest and disability. One connection is between a Native American boy whose home didn't have a phone line until one was installed for network access and his mentor, an engineer in Scotland. Both are deaf. Another DO-IT scholar, a blind girl, has received a full, four-year scholarship from NASA based in part upon her efforts in the NSF program.

At the other end of the spectrum are much broader access-building initiatives. "Rural datafication" is an NSF effort to establish on-line communities among groups with common interests -- such as agriculture, health and K-12 education -- across rural parts of the upper Midwest.

The impact of NSFNET on the computer and communications industry has been equally dramatic. Prior to 1986 only a small handful of networks existed within the research and education community. The introduction of NSFNET stimulated the rapid deployment of regional networks throughout all 50 states and a need for higher speed connectivity. In turn, this stimulated development of switches and routers to accommodate the increasing demands of the networks. Cisco, Proteon, and Wellfleet, as examples, have developed a major router technology within the United States. What has arisen from this activity, with clear market potential, is a new business area of data communications networks. I believe the development fostered by NSF is a major reason why the U.S. leads the world in this area, both in technology and deployment.

A major item on the agenda of the U.S. Government is the National Information Infrastructure. Worldwide, as of April, 1994 the Interagency Internet consisted of more than 30,000 connected IP networks, more than 18,000 of which were in the U.S. There are reported to be million of users on the Internet connected via 1.3 million hosts in 72 countries. Roughly half of this activity is in the U.S. The single most important stimulant to producing this dramatic growth in network communications and its high public profile has been the Interagency Internet program and the associated NSFNET program. In response to this growth of users, the market has responded by the appearance of a variety of commercial providers who offer Internet access at competitive prices.

NSF has recently taken the first steps in what will be an orderly transition from the current NSFNET architecture to a new arrangement that, while advancing both, will distinguish more clearly between cutting-edge experimental efforts and widely used public "commodity" networks.

With transmission capability of 45 million bits of data per second, the current NSFNET backbone makes the connections over commercially leased lines by linking 19 sites -- called nodes -- throughout the U.S. At these sites, mid-level or regional networks are attached and thus interconnected. With partial support from NSF, these regional networks, in turn, reach out to -- and from -- thousands of local networks at schools, universities, libraries, research laboratories, government facilities and supporting commercial organizations.

The NSFNET program long ago outgrew its initial vision: to provide broadband access to NSF's supercomputing centers for researchers with data- or algorithm-intensive projects. The program now supports not only the expanding backbone services, but also directory and information services, operations for regional networks, and connections for universities.

The new architecture calls for the regional networks to secure backbone services from commercial network providers, with a four-year phasing out of NSF's backbone funding for the regionals. Our first priority, as we make this transition, is to insure that the new infrastructure is robust and dependable before we phase out our support for what has become a legendary national backbone service among the networking community.
The transition recognizes the changing nature of the networking marketplace. In 1987 high-bandwidth Internet services were not available commercially. There were neither suppliers nor market. NSF commissioned a high-bandwidth backbone; and, in the seven years since, a number of comparable commercial services have arisen -- catalyzed in part by the market generated by the NSFNET backbone.

The new architecture will allow the research and education communities to take advantage of the excellent services available from commercial suppliers while it provides the interconnectivity that's essential for a vibrant national infrastructure. NSF will continue to emphasize advances in networking and their applications in FY 1995. Specifically, we are introducing an experimental very high-speed bacbone network service (or vBNS), including links to appropriate sites. Our plan is to promote use of this experimental system for high speed applications.

SUPERCOMPUTER CENTERS

Originally established to provide access to researchers in universities, the four NSF Supercomputing Centers have become recognized leaders in the national HPCC effort. By providing reliable high performance computing services, they influence a broad range of scientific disciplines far beyond the academic research community. They train students and scientists in the effective use of HPCC systems. They are involved in the testing of the next generation communications technologies as part of the gigabit testbeds, including developing tools for ease of network navigation. As part of their activities, the Centers develop and freely distribute tools and software for high performance computing, support Grand and National Challenge research efforts, and participate in educational enrichment activities for the K-12 levels.

The Centers pioneered partnerships with the private sector to apply HPCC technologies and practices to specific design and manufacturing problems. They also work with HPCC vendors to provide early feedback from a broad community of users and applications. The placement of Cray Research's new entry into the parallel processing market, the T3D, at the Pittsburgh Supercomputer Center in the fall of 1993 marks the first time a new Cray Research system has been placed at an NSF Supercomputer Center. In a similar vein, IBM chose to place an early version of its first major commercial entry into the scalable parallel computing market at the Cornell University Theory Center. These are significant milestones in furthering academic-industrial collaborations.

To accelerate the potential of HPC in industry, the four Supercomputer Centers currently maintain alliances and partnerships with 88 companies from the petrochemical, computer, pharmaceutical, automotive, chemical, communications and many other U.S. industries. To highlight the innovations that can be achieved in industry,recently the National Center for Supercomputing Application awarded "Top Corporate Honors" to FMC Corporation for its use of high performance computing technology to enhance research into the control of insects and weeds, as well as the optimization of machinery, vehicles and chemical processing plants.

With help from the supercomputer center, FMC had the appropriate technology to perform dozens of complex modeling and simulation projects to improve its current products and contribute to the development of new ones. FMC produced unique three-dimensional models of an enzyme and a receptor protein, both used to design new compounds for pest control. With the ability to visualize and test their ideas in simulation before committing FMC's resources to synthesis, the company's researchers achieve significant savings.. The technology developed by FMC scientists and engineers is also being used by a range of FMC businesses -- including agricultural and industrial chemicals, energy equipment, food machinery, airline equipment and defense equipment. The bottom line: high-performance computing capabilities are increasing the efficiency of processes and contributing to cost savings in unanticipated industrial enterprises.

The NSF Metacenter was organized to facilitate effective collaboration, technical progress, and interoperability among the four NSF Supercomputer Centers. To encourage further extension and collaborative integration of Metacenter activities with those of other providers and users, NSF in FY 1994 initiated support of Metacenter Regional Affiliates (MRAs). MRAs are intended to serve local and regional education, training, research and industrial outreach needs of multiple audiences. In FY 1995, NSF expects to broaden the impact of this program by increasing the number of awards from six to twelve.

In addition, in FY 1995 access to high performance computing services will be enhanced as the supercomputer centers increase both the number of processors on scalable parallel systems and their software activities.

RESEARCH CENTERS AND INFRASTRUCTURE

Through its centers and other research infrastructure activities, NSF's HPCC program has fostered important discoveries and developments that strengthen not only the underlying science, but also a growing computer and communications vendor industry, and user industries. The interactions of these centers and their external advisory and oversight committees provide industry with early knowledge of research results, leading to collaborative efforts to apply those results quickly to appropriate industry problems.

Earlier I mentioned the development of the High-Performance FORTRAN language specification. This and the Message Passing Interface specification will likely guide industry to develop more portable and flexible tools.

Several notable accomplishments have come from other centers' collaborations. A comprehensive light-reflection model for computer graphics now distributed nationally is used for visualization of data in many science applications. Results from research on compilation for parallel and vector computers have been applied by Convex Computer Corporation to its newest products. Joint research resulted in a Bellcore product for better routing of phone calls.

Another way NSF strengthens the infrastructure is by helping universities acquire scalable parallel computing systems and related equipment to support HPCC-related research and education. In FY 1995, this program will be expanded to meet some of the growing research infrastructure and instrumentation needs of the IITA component of the overall HPCC program.

Parallel processing is the focus of much research because it holds promise for immensely improving the ability to solve complex algorithms and to evaluate quickly vast amounts of data. NSF has placed scalable parallel computing systems in twenty academic computer science and engineering departments to enable HPCC research and education. This will enable researchers and students to conduct intensive research on systems that they can control and model.

In the software arena, NSF, in collaboration with the Department of Energy, has supported NetLib, a virtual warehouse of scientific software and datasets, which responded -- automatically -- to more than one million remotely logged requests since it was established eight years ago.

For the past two years, a project called Bioscan has made the processing power of high-performance computing remotely available to biologists whose research requires the comparing of extensive sequences of proteins and genes.

In FY 1995, algorithm and software development for scalable parallel computing systems will be continued, and additional entry-level scalable parallel systems will be provided to universities.

GRAND CHALLENGE APPLICATION GROUPS

The so-called grand challenge applications are exactly that: grand. Just try to predict the weather, design drugs, or analyze the environment and you come up against massive computational requirements.

Grand challenges are defined as "fundamental problems in science and engineering with broad economic and scientific impact, whose solutions require the application of high-performance computing." By their very nature, these problems are beyond the scope of a single researcher, a single scientific discipline, and, sometimes, a single supercomputer. That's why the Grand Challenges activity encourages an integrated, multidisciplinary approach to solutions.

The grand challenge teams weave computation and communication along with theory and experiment in the scientific process. In particular, they show how multidisciplinary teams at geographically distributed sites can effectively approach the complex problems facing science and engineering as we move to the next century. The National Science Foundation, together with the Advanced Research Projects Agency, has launched 16 grand challenge projects.

Projects are selected based on factors including the potential impact in a critical area of science or engineering,potential applicability of computational techniques and systems software, participation of accomplished researchers in multidisciplinary groups, and training opportunities for students. Awards include studies of geophysical fluid dynamics, ground motion in earthquakes, the formation of the universe, bimolecular modelling, human joint mechanics, and large-scale ecosystems.

One grand challenge project is trying to solve a problem basic to all of them: how to coordinate the input and output (I/O) of massive amounts of data between the assembled machines, in this case systems with more than 500 "computational nodes" each. Led by California Institute of

84-115 O - 94 - 2

Technology computer scientist Paul Messina, the researchers will explore advanced I/O patterns and prototypes by focusing on problems of fluid dynamics, chemistry, astronomy, neuroscience, and materials-processing plasmas. These areas were selected because they cover a wide range of I/O needs and behavior, including reading data from high-speed instruments, interactive and irregular data flows, and "pipelining" of data from distributed computing resources on high-speed networks. The project's leader expects it to provide general I/O methods for "sustained TERAFLOPS computers," machines able to handle a trillion floating-point operations per second, as these machines that become available in coming years.

Grand challenge applications have already borne fruit. For example, software developed to support access to the digital library system for unprocessed telescopic data has been distributed to the astronomy community, which has begun to use and evaluate it.

Between four and six additional grand challenge applications groups will be initiated in FY 1995.

DISCIPLINARY HPCC RESEARCH

Many HPCC-related activities are focused on specific disciplines, supporting the development of new knowledge, increasing the broad understanding of high performance computing systems, and training people in the discipline to apply HPCC technology effectively. Among the more than 70 disciplinary research programs are ones advancing areas such as computing systems and components, software systems and algorithms, very high speed networks and optical systems, human-machine interaction and information access, biological sciences, engineering, geosciences, computational mathematics, physical sciences, and social, behavioral and economic sciences.

HPCC technology is an important aspect of the research programs in essentially all of the directorates at NSF. While HPCC is primarily an enabling tool for these other disciplines, there is a widening intellectual integration taking place as these efforts demonstrate the effectiveness of the technology and expand the user base.Example activities include:

- Researchers in the biological sciences have used the Internet to control an electron microscope remotely, thus demonstrating an approach with potential industrial implications.
- The Engineering Directorate supports fundamental research in optical devices that hold promise for very fast switches; these could improve communications networks.
- In the Geosciences, Antarctic science facilities are now connected to the Internet allowing for high speed data transmission of global data and images from telescopes, and making possible general interactive "tele-science" from remote experimental centers.
- A number of new mathematical algorithms broaden the applicability of emerging parallel systems to modeling complex materials and biological molecules.
- Finally, artificial intelligence simulations coupled with experimental laboratory studies have led to better understandings of human social and economic interaction.

HPCC-supported disciplinary efforts, such as optimization algorithms and programs, have fueled progress beyond the researchers' realm.

- In medicine, algorithms based on linear programming were found to be very effective in analyzing x-ray images to diagnose breast cancer. Also, virtual environments are being applied now in medical science, creating new areas of study for pre-operative analysis and for training of medical personnel without the use of expensive "real" procedures.

- In national defense, fleet composition was determined according to diverse U.S. Navy requirements to guarantee maximum responsiveness to a variety of military situations.

Industrial participation in disciplinary activities has also led to major accomplishments.

- For example, gigabit testbed research conducted jointly with industry participants, including baby Bells, Bellcore, IBM, MCI, AT&T, CRAY Research, Pacific Telesis, US West and GTE, has led to experimental high speed (622 Mb/s, 1.2 Gb/s, and 2.4 Gbps) network facilities.

- A software tool, the Rapid Interconnect Circuit Evaluator (RICE) was developed under a research project funded jointly between NSF and industrial sources. In use at a number of companies -- including Texas Instruments (TI), IBM, Intel, LSI Logic and Motorola -- it has reduced the time of an important design evaluation from three days to 20 minutes while enhancing accuracy.

INFORMATION INFRASTRUCTURE TECHNOLOGY AND APPLICATIONS

The proposed National Information Infrastructure will radically change the way Americans generate, transmit, access, and use information. From a schoolchild writing a report on biodiversity, to the librarian who helps the child gather the information from a digital library, to the ecologist who generates the primary data, to the child's parents double-checking with the teacher about the homework assignment, Americans will use and be enriched by the NII daily.

In response to the NII initiative of the White House, the Information Infrastructure Technology and Applications (IITA) component was added to the federal HPCC program. The National Science Foundation chaired a multi-agency IITA task group in August 1993 to lay the foundation for an IITA program that will demonstrate solutions to information-intensive national problems, such as health care, manufacturing, and lifelong education. This component will advance intelligent system interfaces, real environments augmented with virtual environments, image understanding, language and speech understanding, intelligent agents aiding humans "in the loop," and next generation data and object bases for electronic libraries and commerce. These efforts will be coupled with a vigorous program of testbed experimentation to ensure the continued leadership of the United States in critical information processing technologies.

NSF is also working with other federal agencies to strengthen the HPCC technology base, broaden the markets for these technologies, and accelerate industry development of the National Information Infrastructure. This will be accomplished in part by accelerating the development of readily-accessible, widely used, large-scale applications with significant economic and social benefit

The development and implementation of an NII is not a matter of technology alone. That technology is based on knowledge creation. It requires expertise for its meaningful use, tools and interfaces for specific application domains, policies for public use of services, and, last but not least, training of its potential users.

NSF is particularly qualified to make important contributions to the research efforts to create the technologies and to provide the educational foundations on which the NII will be built. NSF's relevance to this mission derives from several different factors, all of which must be integrated in order to achieve the NII vision. First, NSF supports scientific and engineering research across the spectrum of disciplines. Second, NSF, through its NSFNET program, has built an infrastructure for bringing people with a wide spectrum of interests and expertise together through networking and other facilities. Third, NSF is involved with all aspects of the educational enterprise, from graduate education to K-12, from teacher training to systemic change. Fourth, NSF built partnerships between the inventors, producers, educators, and users of NII technologies. Fifth, NSF has extensive experience in innovative, effective collaboration with industry.

These qualifications are critical as NSF addresses the IITA program component holistically and in integrating efforts in four broad topic areas: information Infrastructure services; intelligent interfaces; system development and support environments; and National Challenges. In FY 1994, NSF initiated a program in digital libraries research jointly supported by NSF, ARPA and NASA; and National Challenge programs in civil infrastructure, manufacturing, and health care systems Delivery. Research in Virtual Prototyping in support of the Manufacturing initiative will begin. The Networking Infrastructure for Education program was also initiated in FY 1994. This latter program responds to the National Challenge in Education and is designed to accelerate the adoption of advanced networking technologies in support of science and mathematics education.

In FY 1995, support for all IITA activities initiated in FY 1994 will expand. In addition, pilot projects and testbeds to continue to assess applications of distributed disciplinary data bases and remote access to laboratory instruments will be increased. IITA research activities will include intelligent systems and interfaces, multi-media computing (including virtual reality), information storage, access and query.

EDUCATION AND TRAINING

The NSF HPCC Program also supports activities directly targeted at education and training. These are in three general categories: K-12, undergraduate, and graduate.

The K-12 activities include four pilot network testbeds for education involving federal, state, local school districts and industry partnerships; the Superquest program -- a competition for teams of high school students working on a high performance computing project; and the Global Schoolhouse (GSH). The latter program began with a demonstration of actual class room use of networking and collaboration technology during National Science and Technology Week in April, 1993. Schools in California, Tennessee, Virginia and the United Kingdom took part. The GSH has evolved into a catalyst for the introduction of new networking technologies such as CU-SeeMe Desk Top Conferencing software which is being developed and distributed by Cornell University.

The undergraduate activities include MOSIS (Metal Oxide Semiconductor Implementation Service), support for the undergraduate infrastructure, and a program to support research experiences for undergraduates. The MOSIS program has had a major impact on industry as it trains a new generation of computer chip designers in state-of-the-art design and manufacturing of high performance VLSI computer chips. The MOSIS service has been used in 175 universities in 47 states to train thousands of students each year.

Graduate and postdoctoral training programs are also supported as part of most NSF HPCC Program activities. In addition, approximately eighty postdoctoral research associateships in HPCC have been awarded.

Specific program activities have been designed to broaden the participation of minority institutions and underrepresented groups in high performance computing and communication areas.

PROCESS TO ESTABLISH GOALS AND OBJECTIVES AND EVALUATE PROGRESS

NSF continually reviews, refines and, when necessary, refocuses the overarching long- term goals and objectives of its HPCC Program. At the broadest policy level, the National Science Board, interagency committees and subcommittees, ad hoc panels and committees commissioned to study and recommend program activities, and external advisory committees are the principal contributors to the formulation of the broad long- term goals. In each technical area, research agendas are set, programs are implemented, and evaluations are made through external peer review , technical planning workshops, committees of visitors technical oversight teams, review of final project reports, and input from the researchers and users.

Evaluation of progress and development of long range plans in each class of HPCC activities is integrated into a continuing overall review of proposed new work. Some examples illustrate the nature and extent of this process.

- The High Performance Computing and Communications Information Technology subcommittee, an interagency group, coordinates efforts across the federal government.

  The Federal Networking Council (FNC) is composed of representatives from federal agencies that have requirements for networking in support of research and education and for advancing inter-agency Internet. This group, under NSF leadership, helps coordinate activities in such areas as policy, security, and engineering and operations. The FNC is advised by an external committee with representatives from industry, academia, libraries, and general educational institutions.

- The CISE Directorate Advisory Committee provides policy guidance to CISE.

- The Blue Ribbon Panel for High Performance Computing commissioned by the National Science Board, a group of fourteen scientists from universities and industry, examined the present state of high performance computing and considered the breadth of contributions, barriers to, and opportunities for, high performance computing. The panel provided recommendations to help guide the development of NSF's participation in supercomputing and its relation to the federal interagency HPCC Program.

- Committees of Visitors provide feedback to the Foundation on the quality of the internal management of each program, and on how well the programs have met their goals.

- A series of Workshops on Digital Libraries led to the development of a source book on digital libraries and eventually the joint initiative on digital libraries funded by NSF, ARPA and NASA.

PARTNERSHIPS WITH OTHER AGENCIES

Coordination of NSF HPCC program activities with other government agencies is an essential feature of the NSF program. This includes common and coordinated planning processes, joint sponsorship of large scale research and infrastructure projects, scientific conferences and workshops, and development of shared resources. Several examples show the extent and nature of the variety of collaborative arrangements.

- From the key Memorandum of Understanding with DARPA in 1985 to expand ARPANET for NSF sites, to the establishment of the Federal Networking Council by the NSF Chair of the FCCSET Subcommittee on Networking, NSF has worked with other agencies to coordinate plans and share facilities where practical.

- The NSFNET Connections Program has worked closely with the National Library of Medicine (NLM), making numerous awards using the NSFNET program process and NLM funding to enable health science centers to be connected to the U.S. Internet.

- NSF encourages partnerships between other agencies and the Supercomputer Centers, some of which are at the agency-to-agency level, and others involve an agency directly with a center. In FY 1993, partners include Los Alamos National Laboratory, Livermore National Laboratory, Sandia National Laboratory, Oak Ridge National Laboratory, Argonne National Laboratory; the Maui High Performance Computing Center (U.S. Air Force); NASA Headquarters; NASA Research Centers; U.S. Geological Survey; U.S. Naval Research Centers; and ARPA.

  Many of these multi-agency projects extend over multiple years. FY 1994 funding from the agencies associated with these projects, including NSF, is expected to exceed $12 million.

The following skeletal listing suggests the diversity of mutual activities and interests that have been coordinated:

- Advanced Research Projects Administration - HPC Early Implementation Sites, HPC Consortium, Grand Challenge Applications Groups, Gigabit testbeds, Digital Libraries, Software Technology Research, VLSI Fabrication (MOSIS), STC

- National Aeronautics and Space Administration - Network Connections, Digital Libraries, Caltech Consortium, Software Technology, HPC Performance Evaluation

- National Library of Medicine - Network Connections, HPC Access, Biomedical Applications, Testbeds

- National Institute of Science and Technology - Networking Research, Performance Evaluation

- Department of Energy - Network Connections, Cal Tech Consortium

- Department of Education - Educational Networking Testbeds, and

- Department of Agriculture - Network Connections.

PROGRAM SUMMARY

In summary, the NSF HPCC Program greatly enhances the conduct of science and engineering by enabling the solution of more complex problems, by fostering more frequent collaboration across disciplines, by providing shared access to large collections of data and information, and by enabling the simulation of complex models and processes.

In FY 1995, NSF is requesting support for its HPCC program at a level of $329 million, including about $51 million for the IITA component. The increase of roughly 23 percent over the FY 1994 level will support a balanced set of activities across the original four program components and expand further the IITA effort.

NSF will continue to emphasize networking and its applications in FY 1995 Specifically, transition of the existing backbone to vBNS will be completed and very high speed links will be added to sites with applications with very high speed requirements. In addition, the power of and access to high performance computing services will be continued by increasing the number of processors on scalable parallel systems and software activities at the NSF centers, expanding metacenter activities, and the Metacenter Regional Affiliates program. Additional entry level scalable parallel systems will be provided to universities, additional grand challenge applications Groups will be supported, and algorithm and software development for scalable parallel computing systems will be continued.

Support for the federal National Information Infrastructure initiative will expand under the HPCCIT-coordinated component of the HPCC program. Networking connections for the user community will increase, with special attention to educational activities. An interdisciplinary interagency research program in digital libraries will be expanded, pilot projects and testbeds to assess applications of distributed database and remote access to laboratory instruments will be continued, and activities which support the integration of advanced computing and communications systems for health care delivery, advanced manufacturing, civil infrastructure, and government information dissemination will be increased. IITA research activities will include intelligent systems and interfaces, multimedia computing, including virtual reality, information storage, access and query. An infrastructure program to support National Challenge Applications will be initiated.

Mr. Chairman, I believe that many of the NSF accomplishments and plans that I have sketched suggest how close interagency, academic-industrial partnerships efforts have contributed to making our program more successful.

Mr. BOUCHER. Thank you very much, Dr. Ciment.

Mr. Holcomb.

Mr. HOLCOMB. Mr. Chairman, it is a pleasure for me to appear before you today to discuss NASA's fiscal year 1995 funding request for high performance computing and communications.

High performance computing is vital to NASA's progress in science and engineering and is essential to support many of NASA's programs. While high performance computing and communications is a very small percentage of NASA's budget, it has a very significant impact on our agency mission.

Computation structural analysis, for example, was used to model and redesign the shuttle solid rocket booster joint. In Earth and Space Sciences, supercomputers have enabled NASA to make significant strides in weather and climate simulation, in stellar and galactic evolution simulation, and in solid earth modeling.

A major mission of NASA is to provide advanced technology and facilities to ensure U.S. leadership in aviation. The U.S. aeronautics industry contributed over $23 billion positive balance of trade in 1993, down from $29 billion in 1991. The health of this industry is threatened by declining market share.

High performance computing is vital to the future competitiveness of the U.S. aerospace industry. Rapid, accurate prediction of resistance of air flowing over an airplane, or drag, is fundamental to aircraft design. Today's simulations do not accurately predict drag. High performance computing will enable inexpensive, accurate simulations of aircraft drag. The benefits are superior aircraft design, reduced certification costs, and improved reliability, and, most importantly, a reduction in the overall design process time. For example, high performance computing could reduce the cost from $1 billion to $700 million and the design process from five years to three years for a typical aircraft engine.

Another major mission of NASA is to expand our knowledge and understanding of scientific phenomena. The most striking example of scientific challenge with significant implications for society is increasing our understanding of the climate and determining the extent, causes, and consequences of global climate change. However, our current capability to accurately model the Earth's environment is primitive. Supercomputers are essential to advance model simulation, resolution, and sophistication. Today's models are just beginning to couple atmosphere and ocean models together. Advances in high performance computing are necessary to accurately couple these models and describe the physics, to understand heating and cooling effects, effects of aerosols and water vapor. Modeling biological and human induced effects on the global environment will require even greater increases in supercomputer power.

High performance computing is critical to the development of the National Information Infrastructure. Our goal is an infrastructure which can be used by all Americans, not just leading edge scientists and engineers. We believe this infrastructure will help create jobs, spur growth, and foster U.S. technological leadership, reduce health care costs, deliver higher quality and lower cost government products and services, and, most importantly, I think, prepare our children for the information-rich workplace of the 21st century.

Let me highlight several accomplishments of our program. NASA and DOE joint efforts have accelerated the availability of asynchronous transfer mode, synchronous optical network, or ATM/SONET technology, often called fast packet switched cell relay. Because of Federal investment, this technology is now being deployed nationwide at 45 megabits per second commercially.

In addition, ground terminals under development jointly by NASA and ARPA have been tested at 622 million bits per second speeds through the advanced communications technology satellite in readiness for combined ground fiber and satellite network experiments planned for fiscal year 1995.

To accelerate the process of technology transition, NASA has expanded its use of cooperative research agreements and formation of consortia with academic and industrial partners including hardware suppliers and, most importantly, as Dr. Lindberg indicated, third party software vendors.

Recently NASA selected its first cooperative research consortium headed by the International Business Machine Corporation to carry out research and test parallel computer systems. The consortium members are IBM, Boeing, Rensselaer Polytechnic, Lockheed Missiles and Space, Centric Engineering, Intelligent Aerodynamics, and Rice University. NASA Ames Research Center will be a principal computation site for the new program. Project scientists will use the new IBM SP-2 parallel computer system.

NASA has pioneered the development of widely used and accepted performance evaluation software for applications on parallel computers, the numerical aerodynamics simulation parallel benchmarks. Performance on these benchmarks for the IBM SP-2 is now exceeding performance on the best of traditional vector supercomputers.

Thank you, Mr. Chairman.

[The prepared statement of Mr. Holcomb follows:]

Hold for Release Until
Presented by Witness

National Aeronautics and
Space Administration

May 10, 1994

Subcommittee on Science

Committee on Science, Space and Technology

U.S. House of Representatives

Statement by:

Lee B. Holcomb
Director, High Performance Computing
and Communications
Office of Aeronautics

103rd Congress

HOLD FOR RELEASE UNTIL
PRESENTED BY WITNESS
MAY 10, 1994

STATEMENT
of
Mr Lee B. Holcomb

Director, High Performance Computing and Communications

NATIONAL AERONAUTICS AND SPACE ADMINISTRATION

before the

Subcommittee on Science
Committee on Science, Space and Technology
House of Representatives

Introduction

It is a pleasure for me to appear before you today to discuss NASA's FY 1995 Funding Request for the High Performance Computing and Communications (HPCC) Program.

High-performance computing is vital to NASA's progress in science and engineering, and is essential to support many of NASA's programs in all phases from hypothesis formulation and conceptual design to analysis and missin operations. While the High Performance Computing and Communications Program is a small percentage of the NASA budget, it has a significant impact on our Agency's mission. For example, one application of high-performance computing, computational flow analysis, was utilized in the redesign of the space shuttle main engine liquid oxygen post assembly. Computational structural analysis, another application of high-performance computing, was used to model and redesign the shuttle solid rocket booster (SRB) joint. In Earth and space sciences, supercomputers have enabled NASA to make significant strides in weather/climate simulation, in stellar/galactic evolution simulation, and solid earth modeling.

A major mission of NASA is to provide advanced technology and facilities to ensure U.S. leadership in aviation. The U.S. aeronautics industry contributed over $23.5 billion positive balance of trade in 1993, down from $29 billion in 1991, and the health of this industry is threatened by declining market share. High-performance computing is critical to the future competitiveness of the U.S. aerospace industry. Rapid, accurate prediction of the resistance caused by air flowing over an airplane, or drag, is fundamental to aircraft design. Today's simulations do not accurately predict drag. High-performance computing will enable inexpensive, accurate, 15-minute simulations of aircraft drag. The benefits are superior aircraft design, reduced certification costs, improved reliability and most importantly, a reduction in the design process time. High performance computing could reduce the design cost from one billion dollars to 700 million dollars and the design process time from five years to three years for a typical aircraft engine.

Another major mission of NASA is to expand our knowledge and understanding of scientific phenomena. The most striking example of a scientific challenge with significant implications for society is increasing our understanding of the climate and determining the extent, causes and consequences of global climate change. However, our current capability to accurately model the Earth's environment is primitive. Supercomputers are essential to advance model resolution and sophistication. While today's models are just beginning to couple atmosphere and ocean models, advances in high-performance computing are necessary to accurately couple these models and to better understand the heating and cooling effects of clouds, aerosols and water vapor. Modeling biological and human-induced effects on the global environment will require even greater increases in supercomputing power.

We have evaluated our supercomputing needs with respect to the current HPCC Program goal of achieving sustained computational speeds of a trillion floating point operations per second by 1997. We concluded that computational speeds 1000 times greater that the current HPCC Program goal are required in order to fully meet important aeronautics and space science goals. We must begin now to plant the seeds for the post-1997 generation of supercomputers, and our planned program does that.

The supercomputing industry represents only 1% of the $150 billion computer industry. Supercomputer manufacturers are finding it increasingly difficult to support the rising cost of developing the next generation of supercomputers. National investment in high-performance computing must be viewed in a larger framework of its total benefit to the U.S. economy. The effective use of supercomputing in industrial design and manufacturing could return as much as several hundred billion dollars to the U.S. economy. We cannot afford to slow the pace of advancement and the use of supercomputers. The Administration's attention and commitment to high-performance computing and networking is a recognition of this technology's importance to our country.

The development of the National Information Infrastructure will provide the means by which the United States can achieve a broad range of economic and social goals. An infrastructure which can be used by all Americans, not just scientists and engineers, is our goal. This infrastructure will bring the following benefits to entrepreneurs, factory workers, doctors, teachers, Federal employees, and citizens:

- Create jobs, spur growth, and foster U.S. technological leadership.
- Reduce health care costs while increasing the quality of service in underserved areas.
- Deliver higher-quality, lower-cost government services.
- Prepare our children for the information-rich workplace of the 21st century.

Federal Program Goals

The primary goal of the Federal High Performance Computing and Communications (HPCC) Program is to accelerate the development of high-performance computers and networks and the use of these resources in the Federal government and throughout the American economy. This includes scalable high-performance computing and computer communications networks and advanced software which are critical components of a new National Information Infrastructure.

The Federal HPCC Program is planned, funded, and executed through the close cooperation of Federal agencies and laboratories, private industry, and academia. These efforts are directed toward ensuring that, to the greatest extent possible, the Federal HPCC Program meets the needs of all communities involved and that the results of this Program are brought into the research and educational communities and into the commercial marketplace as rapidly as possible.

The goals of the Federal HPCC Program are to:

- Extend U.S. technological leadership in high-performance computing and computer communications.
- Provide wide dissemination and application of the technologies both to speed the pace of innovation and to improve the nation's economic competitiveness, national security, education, health care, and the global environment.

- Provide a key part of the foundation of the National Information Infrastructure and demonstrate selected Information Infrastructure Applications.

The Federal HPCC Program is fully supportive of and coordinated with the Administration's plan for the National Information Infrastructure (NII). The Federal HPCC Program and its participating agencies plan to provide the basic research and technical development to support the implementation of the NII. To this end, several strategic and programmatic modifications have been made to the Federal HPCC Program. The most significant of these is the inclusion of a new program component called Information Infrastructure Technology and Applications (IITA). It consists of research and development to enable the integration of critical information systems and the application of these systems to "National Challenge" application areas (i.e., civil infrastructure, digital libraries, remote sensing databases, education and lifelong learning, the environment, health care, and manufacturing). The IITA component will enable the extension of critical computing technologies and the development of National Challenge applications that have immediate and direct impact on critical information systems which affect every individual in the nation.

Interagency Roles and Collaborations

The High Performance Computing and Communications Program is implemented as a partnership among federal agencies and other organizations. Program planning is provided by the High Performance Computing, Communications, and Information Technology (HPCCIT) Subcommittee of the Committee on Information and Communication Research and Development (CIC) under the direction of the Office of Science and Technology Policy (OSTP).

NASA leads the HPCCIT Scientific and Engineering Computing Working Group which has the mission to identify, promote, transfer and coordinate software developed under the HPCC Program. The working group meets monthly to coordinate agency HPCC efforts through information exchanges, common development of interagency programs and review individual agency plans and budgets. This working group has sponsored a number of national workshops to stimulate discussion among HPCC software providers and users to develop a national agenda for HPCC software and to promote coordination among agency-sponsored efforts. An HPCC software exchange is also being deployed to provide the infrastructure that encourages software reuse and the sharing of software modules across organizations through an interconnected set of software repositories.

NASA has collaborated with ARPA and NSF in the evaluation of emerging scalable parallel high-performance computer and communication technologies including a joint program to evaluate the role of satellite communications in high-speed networking. The NSF funding for the National Research and Education Network is essential for providing NASA connectivity to academic research institutions. NASA, ARPA and NSF have initiated a joint research program on digital libraries for FY 1993 through FY 1998. NASA, NOAA and EPA have joined to collaborate on the public utilization of national remote sensing data assets over the internet. In the areas of K-12 education, NASA, Department of Education, NSF and DOE have joined forces to accelerate the use of internet in America's schools and to provide educators with network-based educational products.

NASA Program Goals

The goal of NASA's High Performance Computing and Communications Program is to *accelerate the development, application, and transfer of high-performance computing technologies to meet the engineering and science needs of the U.S. aeronautics, Earth and space science communities, and to accelerate the implementation of a National Information Infrastructure.*. This goal is supported by five specific objectives:

1) Develop algorithm and architecture testbeds that are able to fully utilize high-performance computing concepts and increasing end-to-end performance.

2) Develop high-performance computing architectures scalable to sustained TeraFLOPS performance.

3) Demonstrate HPCC technologies on U.S. aeronautics, Earth and space science research problems.

4) Develop services, tools, and interfaces essential to the National Information Infrastructure.

5) Demonstrate pilot programs in remote sensing data access, education and aerospace design and manufacturing.

Program Oversight

NASA HPCC Program evaluations take place at several levels. At the agency level, the NASA Advisory Council (NAC) Ad Hoc Task Force on Supercomputing has completed an external review and report on the NASA HPCC Program. The NAC recommended: (1) the program better recognize and exploit the differing core competencies of industry, academia and NASA so as to define complementary roles that will serve to build a one-team spirit; and (2) solicit from both academia and industry their views on the specific computational aerosciences problems to be considered. In response to these recommendations, NASA significantly improved its involvement of industry and academia in the program planning through workshops, informal writing teams, detailed information exchanges and project plan reviews. To assure the involvement of our key customers, each of the NASA HPCC projects will assign from half to three quarters of their applications efforts to industrial and academic institutions. Specific applications have been re-balanced towards industry and academia needs. New tasks are being added to address affordable computing for industry design environments.

In addition to the NAC, NASA program management policy requires independent annual review of major programs. The HPCC Program independent annual review was recently completed. Appropriate NASA personnel, representatives from other federal agencies, academia and industry participated in this review. Within the HPCC Program, comprehensive reviews are conducted for each of the projects annually. In addition, quarterly reviews of the NASA HPCC program are conducted by program managers, the Associate Administrator for Aeronautics and the Director of the NASA HPCC Office.

Program Accomplishments

NASA has pioneered the development of the most widely accepted performance evaluation software for applications on parallel computers - the Numerical Aerodynamic Simulation (NAS) Parallel Benchmarks. Performance on these benchmarks for best emerging scalable high-performance computers is now exceeding the performance on traditional vector super computer for most of the NAS benchmark suite.

NASA and DOE procurements have accelerated the availability of Asynchronous Transfer Mode/Synchronous Optical Network (ATM/SONET) technology - "fast packet switched" cell relay. This technology is now being deployed nationwide at 45 million bits-per-second speeds by industry. Ground terminals under development for NASA and ARPA have been tested at 622 million bits-per-second speeds through the Advanced Communications Technology Satellite in readiness for combined ground fiber/satellite network experiments planned for FY 1995.

To accelerate the process of technology transition, NASA has expanded its use of cooperative research agreements and formation of consortia with academic and industrial partners. Both horizontal partnerships between NASA and multiple companies in an industrial sector and vertical collaborative teams between hardware suppliers, third-party software vendors and members of applications community are being formed. NASA has recently selected it's first cooperative research consortium, through a competitive selection process. The consortium is headed by International Business Machines (IBM) to carry out research and test parallel computer systems. The consortium members are IBM, Boeing, Rensselaer Polytechnic Institute, Lockheed Missiles and Space,

Centric Engineering, Intelligent Aerodynamics, and Rice University. NASA Ames Research Center will be the principal computational site for the new program. Project scientists will use the new IBM SP-2 parallel computer system.

FY 1995 Request

In FY 1995, NASA's HPCC program is focused in three major areas: integrated, multidisciplinary computational aerospace vehicle design; multidisciplinary modeling and analysis of Earth and space science phenomena; and development of Information Infrastructure Technology and Applications. The FY 1995 budget request is $76.1 million.

NASA plans activities to broaden the reach of the HPCC program and begin the development of a National Information Infrastructure by supporting research and development in education, digital library technology, and public access to Earth and space science data. In education, NASA will increase its focus on developing K-12 education products available over the National Research and Education Network (NREN) and establishing teacher training programs for the utilization of NASA education products. In digital library technology, NASA will advance technology for petabyte scale data storage systems. NASA also will increase efforts to make Earth and space science data more widely available by developing and evaluating prototype digital data bases of images and software that are available over the NREN.

The NASA HPCC Program is structured to contribute to broad Federal efforts while addressing agency-specific computational problems that are beyond projected near term computing capabilities. These computational problems are called *"Grand Challenges."* NASA has selected Grand Challenges in the areas of Computational Aerosciences (CAS) and Earth and Space Sciences (ESS). The Grand Challenge applications have been chosen for their potential and direct impact to NASA, their national importance, and for the technical challenge they give the NASA HPCC Program as a catalyst for reaching Federal goals.

The CAS project will direct its efforts towards continuing the development of multidisciplinary algorithms and advanced software technology in four Grand Challenge areas: high speed civil transport (HSCT), high-performance aircraft, subsonic aircraft, and rotorcraft applications. The ESS project will focus on Grand Challenge teams selected in FY 1993 in the areas of: climate modeling; studying ocean, land and atmosphere dynamics; modeling magnetosphere-solar wind interactions, stellar interiors and surfaces, star/galaxy creation and evolution, and the formation of other cosmological structures; analyzing enormous geophysical databases; and the assimilation of atmospheric data.

Testbeds are a crucial part of this program because they provide a key tool for interdisciplinary research teams to develop and evaluate applications and systems software and to evaluate scalable hardware architectures and peripherals. A key to successful exploitation of massively parallel computing power will be the blending of application-driven and architecture-driven computer systems and software to most effectively meet NASA's needs. Third generation prototype testbeds of 50 to 100 GigaFLOPS are scheduled to become functional in FY 1996.

Summary

The HPCC Program is a critical part of NASA's mission success as well as a main contributor to the overall Federal effort. The HPCC Program is on target to provide a 1,000 fold performance increase in high-performance computing systems and a 100 fold performance increase in computer networking by 1997 over performance levels at the beginning of the program year. These performance increases will be demonstrated on aeronautics and Earth and space science community applications. NASA remains a key participant in the federal program to provide wide dissemination and application of the technologies both to speed the pace of innovation and to improve the nation's economic competitiveness, national security, education, health care, and the global environment. NASA is contributing to accelerating the creation of a National Information Infrastructure with technology and applications driven by NASA missions.

Mr. BOUCHER. Thank you very much, Mr. Holcomb.

Dr. Nelson.

Dr. NELSON. Thank you Mr. Chairman. I appreciate the invitation to testify on the Department of Energy's High Performance Computing and Communication Program.

In my oral remarks I will emphasize just four points; the first of these, the DOE role in the High Performance Computing and Communication Program. The Department's High Performance Computing and Communications Program grows out of our long experience in applying advanced computing data analysis and data communication to our missions in defense, science, energy, the environment, and economic competitiveness.

Today we bring to the HPCC program a strong connection between the development of computing and communications technologies and their early application by sophisticated users in challenging fields. Often these early applications provide feedback to manufacturers or developers on improvements that are needed before the technologies can be widely deployed.

The second point I would like to make is the interagency program. I emphasize, as the other speakers have, that our own program has benefitted greatly from our participation in the interagency HPCC program. The interagency program is broader and deeper than any one agency would achieve. Each agency brings its own skills and missions to the program. Each agency takes advantage of the accomplishments of other participants. The interagency program reduces the risk to each participant by making possible several simultaneous strategies for overcoming those risks. The interagency program has helped to create a critical mass in developing and applying advanced computing and communications technologies.

Essential to achieving these benefits has been the close cooperation among the participating agencies in planning and executing the HPCC program. These agencies have worked together since the early 1980's, and I number the people at this table as among my closest friends. Under the leadership of the Office of Science and Technology Policy and the coordination of the National Coordination Office, I believe the HPCC has become, in effect, a virtual agency.

The third point that I would like to stress is results already occuring from the program. Results from the DOE HPCC program, I believe, are already showing that the investment has been a good one, and I emphasize that many of these results could not have been achieved without the cooperation of the other agencies that we work with and of the private sector.

In science HPCC has been applied by DOE to chemistry, physics, biology, engineering, materials, and the environment, and other areas. In industrial applications, several of our industrial partners through firms and consortia are participating in our program and are already beginning to use the results of the program in their design and manufacture in areas as diverse as chemicals, automotive, pharmaceuticals, and materials. This year, one of our HPCC developments won an R&D 100 award of which we are proud.

In education, the HPCC program has been used in curricula in K-12, college programs, and postgraduate both in developing technologies for delivering education and in using HPCC to stimulate and enrich the educational environment. This year, to show that those results are, in fact, real, we have commissioned assessments, outside assessments, of both our applications area and of our educational area, and the results of those assessments should be ready in the next few months.

The fourth point that I would like to emphasize is our fiscal year 1995 budget request. As is well known to the committee, our budget request for fiscal year 1995 is approximately the same as for 1994, and most of our programs in 1995 will be a continuation of 1994 activities. Our participation in the Information Infrastructure Technology and Applications area will be modest, although we have commissioned a start in the area of energy demand and supply management which offers potential for enormous cost savings and efficiency in energy production and usage.

We hope that we can also leverage some of our other programs in order to begin work in some of the applications areas in IITA and delivering information to the public in environmental remediation and advanced manufacturing.

This concludes my remarks, and afterward I would be pleased to answer any questions you might have.

[The prepared statement of Dr. Nelson follows:]

STATEMENT OF

DAVID B. NELSON

ASSOCIATE DIRECTOR FOR SCIENTIFIC COMPUTING

OFFICE OF ENERGY RESEARCH

DEPARTMENT OF ENERGY

BEFORE THE

COMMITTEE ON SCIENCE, SPACE, AND TECHNOLOGY

SUBCOMMITTEE ON SCIENCE

UNITED STATES HOUSE OF REPRESENTATIVES

MAY 10, 1994

Mr. Chairman and Members of the Subcommittee:

I am pleased to be here today to discuss the High Performance Computing and Communications (HPCC) Program.

The Department of Energy (DOE) and I, personally, are very committed to the HPCC Program. The Department has a long history of involvement in high performance computing and in pioneering many advances in leading edge computing. Since the beginning of the electronic computing age, over 40 years ago, DOE and its predecessor agencies have been at the leading edge of computing. We learned early on to use computing as an important tool in parallel with experiment and with theory to solve scientific research problems. This comprises a discipline now referred to as Computational Science. Comprehensive computational science methods to investigate or perform basic research in energy and matter and for advanced energy applications will continue to require even more significant advances in high performance computing than we have achieved to date and, in fact, more than we will achieve in the next few years in the HPCC Program.

Examples of these research problems and applications under the Department's mission include: the development of energy sources, such as fusion energy; oil reservoir management and enhanced recovery; the development of new materials, such as superconductors, ceramics and metal alloys; understanding biological processes, such as protein structure, the human genetic code, and energy biosciences; virtual prototyping and nuclear safety; understanding global climate change; molecular chemistry technologies for environmental restoration; and finally, probing the fundamental structure of matter in our high energy and nuclear physics programs.

The Department has worked with industry through the years to bring many new supercomputer systems into the marketplace to work on these research problems. Some of the companies that the Department has worked with in this regard are Univac, IBM, Control Data, Cray Research, Intel, Thinking Machines, and Kendall Square. Throughout its history and because of its missions, the Department's involvement in high performance computing has been driven by applications. Because of this, the Department's High Performance Computing research programs have benefitted from expertise of its sophisticated user community. Our partners in industry have likewise benefitted from the close working relationship with, and the feedback from members of this very talented community, who were willing to take the risks associated with using the advanced prototype computer systems which are critical to carry out their research.

The Department, in 1974, also established the nation's first nationwide supercomputer access program to allow energy researchers to have access to the advanced computational tools required to perform their work. To do this, the Department implemented a nationwide high performance communications network, supporting thousands of users across the nation mainly at the national laboratories, but also at universities and industrial sites. Today this network, through the guidance of our user community, has evolved into the Energy Sciences Network (ESNet) -- the first high speed, multi-protocol network in the Internet. This ESNet, together with the National Aeronautics and Space Administration (NASA) Science Internet and the National Science Foundation (NSF) Network, form an important core of the Global Internet, which has become indispensable to the Nation's research community.

The current HPCC Program was developed over the last twelve years. Yes, even though it is only a few years since this Program was formally begun, many government research managers and members of the nation's research community studied and reported on the need for such a program for many years prior to its start up. The requirements, goals, objectives and proposed budgets for the program were the subject of many analyses and reports prior to its implementation. The analyses confirmed the benefits and economic importance of high performance computing and communications to the Nation.

We are now beginning to have even clearer evidence of this importance as initial results of the HPCC Program unfold. Some of the Department's accomplishments in the program to date follow.

- An Operating System for Massively Parallel Computers has been developed which uses only two percent of each processor's memory and operates at peak communication bandwidth. This operating system enabled Sandia National Laboratory researchers to set a 72.9 billion of operations per second record for the widely accepted MPLINPACK benchmark for production parallel computers and enables researchers to execute much larger and more realistic simulations than previously possible.

- Lawrence Berkeley Laboratory has developed Video Audio Transport and Whiteboard capabilities which are currently being used on the ESNet and Internet to support distributed collaboration and desk-top or conference room video conferencing activities. These capabilities are a major success and are increasingly used to videocast scientific meetings, thereby involving more participants in these activities.

- Los Alamos National Laboratory has developed and implemented an 800 megabit per second network Interface which has been installed in the CASA gigabit testbed. This High Performance Parallel Interface technology allows large-scale computers at widely distant locations to exchange information at high data rates, as part of distributed processes.

- Argonne National Laboratory has developed automated differentiation technology software, ADIFOR, which has been used to transform a General Motors Corporation simulation code, LINEFORM, into a design tool for producing dies that will last longer and lead to lower metal stress manufacturing techniques.

- Argonne National Laboratory applied their parallel computing algorithms and software technology to the Allied-Signal Corporation's materials design simulation code, phascan, which is used for designing improved and more environmentally friendly materials for catalytic converters.

- Sandia National Laboratory researchers have implemented a new massively parallel finite element modeling code which can treat problems in electromagnetism and in heat transfer for arbitrary two and three dimensional geometries. This code is being used to design better semiconductor chips and their ceramic packages.

- Sandia researchers have used the supercomputer's massive memory and computing capacities to create detailed, animated three-dimensional images with data collected through magnetic resonance imaging (MRI). These images have been shown to be much more sensitive than conventional mammography for detecting breast cancer. This is the first successful use of MRI in breast imaging. It has produced dramatic results in a small clinical study and has been shown to be significantly more accurate than mammography.

- A new software algorithm to generate computational meshes, called PAVING, has been created. Use of the PAVING algorithm by Ford Motor Company in simulated crash testing has greatly reduced the time which is presently used to generate a mesh. This is extremely important, since current technology requires, for example, six to eight weeks to generate an automobile mesh for a crash simulation and does not allow engineers flexibility they need to accelerate the design process.

- The Lawrence Livermore National Laboratory consortium, the National Storage Laboratory, will soon release a beta version of the software technology needed to manage massive computer data archives in a high performance environment. This collaboration involves several major U.S. industrial partners, including IBM, Ampex, Maximum Strategies, and Open Vision.

- An Adventures in Supercomputing program has been conducted by the Oak Ridge and Sandia National Laboratories and Ames Laboratory. Its primary goal is to capture and cultivate the interest of minority and disadvantaged students, elementary through high school, in science, mathematics, and computing. This program has been co-sponsored by N Cube Corporation and has been extraordinarily successful both in gaining students interests and in helping teachers learn to use information technologies.

- Under the Oak Ridge National Laboratory leadership, an online, hypertext graduate textbook in computational science has been developed and released over the internet. It met an urgent need for a modern textbook in this field. It is currently being used in over a dozen university graduate schools.

- Oak Ridge National Laboratory also leads the Partnership in Computational Sciences (PICS) whose members include Ames Laboratory, Brookhaven National Laboratory, the State University of New York, the University of South Carolina, and Texas A&M University. The PICS consortium is working in several research areas.

The HPCC Program is implemented as a partnership among federal agencies. Leadership for the Program is provided by the Office of Science and Technology Policy. The National Coordination Office for HPCC provides a central focus for program implementation. The Office coordinates the activities of the agencies, other organizations participating in the program, and organizations providing services to it. The Office also provides an interface to Congress, industry, academia, and the public. Dr. Donald A. B. Lindberg, the National Coordination Office Director (and also Director of the National Library of Medicine), reports to Dr. John H. Gibbons, the Assistant to the President for Science and Technology and the Director of the Office of Science and Technology Policy.

The High Performance Computing, Communications, and Information Technology (HPCCIT) Subcommittee of the Committee on Information and Communications of the new National Science and Technology Council handles planning for the HPCC Program. The Subcommittee meets monthly to coordinate agency HPCC Programs through information exchanges, the common development of interagency programs, and the review of individual agency plans and budgets. HPCCIT charters a variety of Working Groups to coordinate activities in specific areas, Task Groups to carry out specific limited duration studies, and Task Forces to make various planning recommendations. In most cases, individual agency representatives are assigned responsibility to lead these groups.

Each year the National Coordination Office also schedules and coordinates meetings to review each agency's budget and program proposals for interagency evaluation and analysis, activities to prepare the multi-agency program plans to articulate the program in the President's budget, activities to prepare multi-agency implementation plans, and public meetings with interested private sector companies, groups, or individuals to get feedback on the program.

These are the major activities which enable the interagency HPCCIT Subcommittee to assess the progress toward meeting the program's goals. My view is that, as a result of this interagency coordinated program development and execution, the HPCC Program is a very balanced program in several ways. There is balance in investment in program components; that is, balanced investment in hardware architectures, in systems software for these systems, in centers and system prototypes for evaluation against real applications, in applications themselves with regard to computational science methods, in software libraries and tools to support these applications, in network research and deployment to access these systems and to support applications collaborations, in basic research to drive creativity and innovation in all areas, and in human resources to provide for a qualified, diverse workforce for the Nation in this critical technology area.

We also have balanced community involvement in the program. This occurs across agencies, but also in another way. Industry is funded more by one agency, the Advanced Research Projects Agency, than others, universities more by the NSF, and laboratories and centers more by the DOE and NASA.

I believe we are also achieving balanced accomplishments, such as those noted above, across HPCC Program components. In addition, we are seeing balance in the application areas being funded in the Grand Challenge collaborations. Grand Challenges are areas of broad national importance that can be significantly advanced by high performance computing.

To select relevant and meritorious Grand Challenges, the Department set up an interagency panel, to evaluate proposals, comprised of its program managers and those of other agencies' participants, chaired by a program director from NASA. The selected projects are balanced across the Department's mission spectrum. They are also cofunded by other laboratory, university, industry, or agency programs. The Department's Grand Challenge research areas, funded by the Office of Scientific Computing, are:

- Computational Chemistry -- to develop parallel chemistry codes important to the study of several critical environmental problems. This project is cofunded by five industrial partners (Allied Signal, Amoco, Dupont, Exxon, and Phillips) and the DOE Chemical Sciences program.

- Computational Structural Biology -- for research in computational methods to understand the components of genomes and to develop a parallel programming environment for structural biology. This project is cofunded by the DOE Office of Health and Environmental Research, the National Science Foundation, J. D. Searles, and the cancer program at the University of Tennessee.

- Mathematical Combustion Modeling -- to develop parallel algorithms for computational fluid dynamics and apply these methods to important combustion models. This project is cofunded with the DOE Applied Mathematics and Chemical Sciences program.

- Quantum Chromodynamics Calculations -- to develop lattice gauge algorithms to study elementary particle physics on massively parallel machines. This project is cofunded by the DOE and NSF high energy physics programs and involves particle physicists from a dozen universities.

- Oil Reservoir Modeling -- to construct efficient parallel algorithms to simulate fluid flow through permeable media. The work is based on reservoir models from the University of Texas that are widely used by the oil industry and is cofunded by DOE Environmental Sciences, the Center for Petroleum and Geosystems Engineering at the University of Texas (sponsored by the major oil companies), Digital Equipment Corporation, NSF, and the State of Texas.

- The Numerical Tokamak Project -- to develop and integrate particle and fluid plasma models on massively parallel machines as part of a multidisciplinary study of tokamak fusion reactors. The project is cofunded by DOE's Fusion Energy program and is an important component of the Energy Policy Act.

- First Principles Simulation of Materials Properties -- to develop parallel techniques and algorithms for materials problems that are beyond the limits of present techniques and guide the discovery of advanced materials. This project is a Partnership in Computational Sciences collaboration involving three DOE laboratories and cofunded by the DOE's program in Materials Science.

- Environmental Remediation: Groundwater Transport -- to improve our capability to model the transport of subsurface contaminants and evaluate alternate remediation strategies through the use of advanced numerical techniques and massively parallel computers. This is a Partnership in Computational Sciences collaboration involving ten partners.

- Global Climate Modeling -- to expand the current theoretical basis for climate dynamics and optimize computer models to best utilize new and improved computer architectures. The current research emphasizes the development of global climate models using massively parallel machines using message passing techniques. This work is cofunded by the DOE Office of Health and Environmental Research as a part of the Global Change Research Program.

These Grand Challenges are very computationally intensive applications which push the envelope in high performance computing and which hold great potential for future research and contribution to the Department's programs. However, these are really just another step along the way to much more comprehensive studies and solutions to these research topics. Full solutions to these problems, and partial solutions to even more difficult or more computationally intensive problems, will require even larger advances in our computational capabilities in the future.

Nevertheless, what may be most important about the HPCC Program today is the fact that high performance computing and communications technologies also lay the technological foundation for the National Information Infrastructure.

The DOE has several mission areas which can benefit greatly from the National Information Infrastructure and which would be beneficial to the Nation. These are: energy demand and supply management, environment, education, manufacturing, oil and gas production, and digital libraries. I will touch on one of these -- energy demand and supply management (primarily electricity and natural gas) -- because I believe this to have enormous potential in energy security and conservation and also in savings in deployment costs for the National Information Infrastructure.

Some studies estimate that electric utilities can deploy the communications infrastructure required for energy demand and supply management for less than it would cost to build future power plants; that is, the energy savings could in fact cover much of the cost. This is extremely significant. Even so, the energy savings due to energy demand and supply management might well offer the most economic and environmental benefits that can be realized

84-115 O - 94 - 3

by the National Information Infrastructure. In Fiscal Year 1995, the Department, in cooperation with the Electric Power Research Institute and the Edison Electric Institute, will have completed a comprehensive study in this regard.

One scenario for wide-spread information infrastructure deployment to homes and businesses is interactive demand-side energy management technology, now being tested by some electric utilities. Utilizing two to five percent of fiber technology capacity, energy demand management technologies may offer sufficient energy savings (especially during periods of peak demand) to electric utilities to warrant their investment in both the technology and its pathway. The Progressive Policy Institute, a Washington-based research group, has studied interactive demand-side energy management and identified it as a credible means of achieving early deployment of the Information Superhighway.

A notable example of this idea is the electric utility, Entergy Corporation's, implementation of a demand-side management system which Entergy plans to deploy ultimately to 442,000 customers in Arkansas, Louisiana and Mississippi. The system will make use of Entergy's 700 miles of fiber optic cable already in the company's service territories. Entergy envisions a direct fiber optic (or hybrid fiber/coaxial cable) hook-up to a residence with the following potential applications for the utilities: (1) demand-side management with continuous, automated feedback; (2) accumulation of customer usage information (i.e., load research); (3) remote reading of meters which could be applicable to electric, gas, water and other devices; (4) automated billing for customers; (5) variable spot pricing; (6) remote turn-on/turn-off capabilities; (7) automated "power-off" indication; (8) automated current diversion; (9) distribution automation; and (10) enhanced customer service. The demand-side management system can identify and control energy usage by individual appliances in existing homes.

As for benefits to the consumer, or user of National Information Infrastructure, Entergy estimates that consumers can also save as much as 20 percent on their electric bills. In addition, deployment costs will be reduced, universal service can be attained, and other economically beneficial National Information Infrastructure applications can be started. The Department's role in this application would be in technology development for broad infrastructure connectivity and interoperability, in gathering and disseminating data on its value, and in creating the forum to facilitate and accelerate the utilities involvement in this area.

Personally, I believe that computing, communications, and information technologies are valuable investments for the Nation in this age of information to maintain our leadership in the global economy. Indeed, the Clinton Administration has placed a high priority in this area. The base HPCC Program and the National Information Infrastructure programs are needed to stimulate leading-edge information technologies, to diffuse the use of advanced computational technologies throughout U.S. industries, and to develop national scale applications for the Nation to benefit our citizens, such as in health care, environment monitoring and remediation, and energy demand and supply management.

The DOE, however, during this past year, has seen its overall budget reduced considerably from the previous year. Inside the Department, the Energy Research budget has been reduced approximately 12 percent. These reductions have been the result of needed efforts to reduce the deficit and the result of changing priorities, especially in the Department, for the post-Cold War era and for environmental remediation. In light of this, the Department's budget for HPCC, although not growing as well as those in the other agencies, has fared much better when taken in relation to other Departmental research budgets.

In summary, Mr. Chairman, I believe that the HPCC Program has proven to be a well-conceived, balanced, and productive program in an area which is important to the nation's economic competitiveness. The Department appreciates your continued support of this program, and requests that you also support the development of the Administration's program to create a National Information Infrastructure.

This concludes my prepared testimony. I would be happy to answer your questions.

Mr. BOUCHER. Thank you, Dr. Nelson.

Mr. Toole.

Mr. TOOLE. Good morning, Mr. Chairman. Thank you very much for the opportunity to appear before your committee and represent ARPA's high performance computing and communications program.

DOD is both a leading user and developer of information technology and is dependent upon information technology for virtually every program that we have throughout the Department. ARPA's principals for success are founded on our R&D experience over the years. Since we felt computing and communications would grow to be pervasive for all individuals, the simple concept of scaling has become very important in all aspects of our program.

We have established five broad goals for the DOD part of the HPCC program which I would like to share with you. First, the HPCC program will develop the high performance computing and communications technologies necessary to meet DOD's most demanding computing needs, both computationally intensive and information intensive. Second, the HPCC program will develop the technology needed to create user-developed software that will be independent of machine architecture. This capability is necessary to avoid costly reprogramming by the user base.

Third, the HPCC program is developing critical expertise in scalable computing such as components, interconnects, computing architecture, and systems software; basically developing a technology base in the computing field.

Fourth, the HPCC program will accelerate the integration of computing and communications. Communications will be an integral part of all computational systems, and the NII, referred to by my colleagues, will require this integration.

Fifth, the HPCC program is striving to create a self-sustaining industrial base that can provide high performance computing and communications systems services, which will meet the needs of the Federal Government, as well as the commercial sector.

As noted in each of our respective testimonies, the level of cooperation between each of the participating agencies resulting in this kind of virtual agency that Dr. Nelson referred to is really unprecedented. As a result, it is important to look across the programs since it is unrealistic to expect any one agency to do everything.

ARPA has taken the lead in developing the scalable computing systems, both hardware and software, that are being used in several of the applications under development by other HPCC participants. ARPA also has a major role in developing the high performance networking technology capable of providing networks with communication band width in excess of a billion bits per second. Examples of these relationships are in my written testimony.

Several of ARPA's R&D programs across ARPA contribute to the whole program. The largest component is a program called high performance computing, ST-19. Key areas include scalable computing, microsystems, scalable software, national scale information enterprise network, and the details also are included in my written testimony.

The innovation and accomplishments of the ARPA programs span a large and diverse sector of projects. Summaries of these projects are available on line through the Internet, so anyone may browse through this project list. Let me give you a couple of examples.

One strategic goal is to bring about a unification across the prevailing models of computation within the scalable computing framework. For example, Cray's new T3D system introduced last summer went from concept to operation in only 26 months and represents a new approach for integrating vector and massive parallel processing technologies. Machines of this class from Intel and many other vendors are now reporting results of significant economic and societal impacts, such as analyzing seismic data to uncover new oil fields in the Gulf of Mexico, simulating large molecular interactions leading to better understanding of the enzyme base cause for Alzheimer's disease, reducing design time, as referenced by Mr. Holcomb, in the aerospace industry, and many, many others.

Scalable software has also been a critical component of our program. The same operating system concepts demonstrated under ARPA sponsorship on very large machines have been adopted by leading manufacturers of work stations as well. Microsoft Windows NT operating system was greatly influenced by the ARPA-developed mock operating system which is also the basis for many of the operating systems running on a variety of scalable parallel machines. The advanced compiler technology developed under ARPA sponsorship at Rice and used by some of my colleagues here has now been adopted by Convex for its latest generation of scalable systems.

Networking, of course, has also been a dominant theme in the program from the earliest days of packet switch to ARPANET to today's Internet with millions of users. ARPA's program in conjunction with the gigabit testbed research with NSF has demonstrated medical, terrain visualization, modeling applications, and gigabit class networks and is demonstrating optical and ATM technology operating over fiber and satellite media.

On the financial side, there are two complementary views of the budget. The detailed descriptions of each of these ARPA products contributing to the Federal program are contained in the recently distributed implementation plan and discussed with your staffs. ARPA's 1995 Federal HPCC request is $357.4 million. I would like to briefly describe some strategic directions of the 1995 program focusing on our key projects.

First, a major increase is planned to support the Administration's National Information Infrastructure. The HPCC program's research investments in advanced computing and networking technologies lay the foundation for this critical national scale enterprise. We are developing a layered architecture for the NII that will ensure high leverage to a common set of end user and application developer services that bridge the network bitways and the National Challenge applications.

Second, we plan to increase the software technology investments. A series of collaborative pilot projects will be established aimed at unifying compiler technology, run time systems, and support for scalable input/output. Software has always been an integral part of the ARPA HPCC program and the program in general, and we currently estimate that approximately 55 percent of our fiscal year

1994 program of high performance computing project is software related.

Third, we plan to increase investments in critical areas of scalable systems. An important variant of scalable systems called embeddable systems will receive increased funding in fiscal year 1995. In addition, the early evaluation component will be reinstated to ensure the earliest possible feedback between the industry and users of new systems. This is being done in conjunction with our other agencies.

Fourth, the networking world must be prepared for a balanced environment of information security, high performance, and advanced services. We plan to increase investment in network scaling issues, information security, and couple these efforts to our IITA components.

Fifth, we plan to apply HPCC to design technology and its experimental applications to computing.

You have asked for a little assessment in terms of how programs are evaluated and priorities and goals are established. ARPA reviews its HPCC efforts at many different levels to ensure consistent evaluation in a very dynamic R&D environment. This is our strength. An annual process of updated accomplishments, milestones, project plans, site visits, project meetings, principal investigator meetings, and regular interactions with industry, industrial groups, academia and other government agencies occur frequently.

The bottom line is that the agency plans, descriptive summaries, and other products produced in the program reflect these types of things such as described in the implementation plan.

In summary, I strongly believe that the Federal HPCC program represents a unique and unprecedented effort across the Federal Government. I am proud to be part of it, and it has been very successful in exceeding its goals.

I would be glad to answer any questions you may have for me, sir.

[The prepared statement of Mr. Toole follows:]

Statement by

JOHN C. TOOLE

**ACTING DIRECTOR,
COMPUTING SYSTEMS TECHNOLOGY OFFICE,
ADVANCED RESEARCH PROJECTS AGENCY**

Before the

SUBCOMMITTEE ON SCIENCE

HOUSE COMMITTEE ON SCIENCE, SPACE, AND TECHNOLOGY

May 10, 1994

**Not for Publication until
Released by the House Science,
Space and Technology Committee**

ARPA High Performance Computing and Communications Program
(May 10, 1994)

INTRODUCTION

Good Morning Mr. Chairman, Members of the Subcommittee, and staff. I am pleased to appear before this subcommittee and describe the ARPA High Performance Computing and Communications (HPCC) program. ARPA, as the central research and development arm of the Department of Defense, has an impressive record of accomplishment in computing and information technology, from the early contributions to timesharing, computer graphics and networking technology to today's program in high performance computing and communications. DoD is both a leading user and developer of information technology. The Defense Department is dependent upon information technology for our surveillance, communication, command and control, and weapon systems; and increasingly to help us train our forces and produce affordable weapon systems.

OVERALL GOALS AND PRINCIPLES

By the mid 1980's, it was clear that the Federal and industrial investments in scalable parallel computing had enormous potential. ARPA's Strategic Computing program, which began in 1983, and other government agency's early investments in parallel computation provided the foundation to conclude that scalable parallel concepts are real; HPCC is strategically important to the US; and scalable systems emerging from industry have great potential. However, an enormous amount of work is still needed to make the concepts mature, reinvent the technology base, and provide results to the increasing number of users across government, industry, and academia.

ARPA's principles for success are founded on our R&D experience over the years in this field. Since we felt computing and communications would grow to be pervasive for all individuals, the simple concept of scaling became very important. Simply stated, technology investments, including hardware, software, and components, are made with the overall idea that subsystems can be reused in multiple configurations, from single workstations to the highest performing computing systems. Unlike just building one huge computer, scaling allows you to productively invest in smaller, yet important, efforts and have a research infrastructure available when you are completed. Although we normally think of scaling up from high performance workstation clusters to massively parallel machines, HPCC is also investing in scaling across multiple media by insuring access to these same networks through wireless, satellite, and hybrid networks.

ARPA has established 5 broad goals for the DoD part of the HPCC program to meet our nation's defense needs. Many of these goals are common with the other Federal agencies which participate in this program, and contribute to the strength of our nation's economic base:

- First, the HPCC program will develop the high performance computing and communications technologies necessary to meet the DoD's most demanding computing needs. These include computation intensive applications as well as information intensive applications with great impact on both Defense as well as the economic well being of the Nation. In addition, we want to replace completely custom built systems with scalable modules wherever possible, thereby reducing cost, improving interoperability, and focusing precious dollars on only those custom subsystems that have the most leverage. The HPCC program has focused on generic enabling technologies, rather than much more limited point solutions as a means to achieve this goal.

- Second, the HPCC program will develop the technology needed to create user-developed software that will be independent of machine architecture. This capability is necessary to avoid costly reprogramming by the users.

- Third, the HPCC program is developing critical expertise in scalable computing areas such as components, interconnects, computing architectures, and systems software. This will allow us to realize distributed computing over heterogeneous platforms, from workstations and clusters to large-scale, high performance systems. In addition, embeddable systems, capable of exploiting the commercial baseline, will not be possible without this expertise.

-Fourth, the HPCC program will accelerate the integration

of computing and communications. Communications will be an integral part of all computational systems, ranging from communications internal to the architecture of high performance systems, to client server systems on local networks and to the use of national resources located on networks across the country. The National Information Infrastructure (NII) will require the integration of computing and communications.

- Fifth, the HPCC program is striving to create a self-sustaining industrial base that can provide high performance computing and communication systems and services which meets the needs of the Federal government as well as the commercial sector.

Each of these goals meets needs that these industries have expressed in a variety of forums, and the HPCC program is stimulating an open, innovative, and competitive process. Across the entire computing and communications sector, we want to be able to rapidly develop HPCC applications which can be used by a variety of users across major platforms to solve critical problems, both for Defense and the commercial sector. As you have heard, the National Program is addressing the opportunity to apply the technologies of HPCC to other major programs.

RELATIONSHIP TO OTHER FEDERAL PROGRAMS

As noted in each of our respective testimonies, the level of cooperation between each of the participating agencies, resulting in a kind of "virtual agency," is unprecedented. As a result, it is important to look across the programs, since it is unrealistic to expect any one program to do everything. However, our dependencies and relationships are important.

The activities of all the participating agencies are coordinated through the National Coordination Office which is under the auspices of the Committee on Information and Communications of the National Science and Technology Council. ARPA has taken the lead in developing the scalable computing systems (both the hardware and systems software) that are being used in several of the applications under development by other HPCC participants. ARPA also has a major role in developing the high performance networking technology, capable of providing networks with communication bandwidth in excess of a billion bits per second.

The technology base in computing and communications must have a mix of investments in very aggressive theory, small scale experiments, mid-scale deployed technology, and industrial partnerships and collaborations. On the research side of this spectrum, we have a special relationship with NSF in many of our programs. We couple to NASA as a large-scale experimental user of our early technology in areas vital to Defense. DOE becomes a major user of our scalable computing systems as well as stressing them for very large problems. NSA's results in encryption is important to our networking and computing program, as well as being a large consumer of scalable parallel systems. NIST's role in standards and working with the industry will enable the transfer of R&D, particularly in the IITA component. NIH, EPA, and NOAA become major pilot users of advanced computing and information technologies are a means to rapidly transition technology.

Numerous examples highlight these relationships, which have been documented in several sources; however, a couple illustrate how synergy develops across the programs:

- The Gigabit testbeds, cofunded by NSF and ARPA, with participation by DOE and other Agencies, have helped focus high performance networking on some specific problems critical to the program.

- ARPA, in its early evaluation work, has joined with NSF, NASA, NIH, and DOE to evaluate prototype systems with real users, and to push the limits of code development, environments, and capabilities for a diverse set of users.

- ARPA has a program working with NSA to deploy a trusted version of the Mach operating system, designed to be a foundation for a secure environment in the future.

OVERVIEW OF PROGRAM

Several of ARPA's R&D programs contribute to HPCC. The largest component of the ARPA request to the Federal Program is ST-19, a project called High Performance Computing. The overall

ARPA program can most easily be understood in the following major categories:

- Scalable Computing - includes the research and development of scalable components and systems, as well as their evaluation in early experimental use. Embeddable Systems is part of this component, which enables scalable components to be used in different ways for a variety of users.

- Microsystems - develops the design tools and prototyping infrastructure to support the R&D of advanced scalable parallel computing components and systems for large-scale systems, embedded systems, and wireless computing systems. Innovative prototyping and small-scale experiments are also supported which leverage microarchitectures advanced packaging, and low energy components.

- Scalable Software - develops technologies for operating systems, programming languages, compilers, tools, and environments to enable the effective use of the new high performance computing technologies.

- National-Scale Information Enterprise - develops the underlying technologies and services to support large, complex, and distributed applications. Applications include health care, design and manufacturing, etc. while services might include privacy and trust mechanisms, remote resource sharing, information security and others. This area builds on previous work in distributed systems, but focuses on the longer term technology research issues needed to realize the National Information Infrastructure. In addition, we intend to work with other programs inside and outside of Defense that are interested in very advanced applications.

-Networking - develops high performance networking technologies and capabilities which permit the technology to scale to a broad class of users. ARPA's role is to work closely with other programs and the industry, focus on the longer term technologies and research issues, and implement an accelerated transition strategy for successful results.

ARPA's role is probably most noted for its R&D in scalable architectures, system software, and networking research. New approaches in these areas have been developed, long before it was generally considered popular. This requires substantial investments from the University levels up through cost-shared arrangements with industry. For example, the fundamental research at universities in scalable shared memory multiprocessors architectures has been the foundation for commercially developed systems. Results from these projects have been used in some of the larger cost-shared efforts with Intel, TERA, Cray Research Inc., Thinking Machines, and many others. In the 1980's, this was very radical thinking, and now every major vendor has recognized the importance of scalable parallel systems. However, it is important to follow through, since this represents an unprecedented reinvention of the field in a very volatile time.

ACCOMPLISHMENTS

The innovations and accomplishments of the ARPA program span a large and diverse number of projects. In FY93, for example, the High Performance Computing Project had approximately 334 distinct projects - 42% from universities, 43% from industry, and 15% from FFRDC's and government laboratories. Summaries of projects are now available through on-line servers, so anyone may browse through the project list. Although we cannot cover all of them, I'd like to give you an idea of some representative examples, which indicate the enormous progress the program has made, plus an indication of the sustained investments and time it takes to realize the gains from the work.

Innovation in scalable computing research, both in hardware and software, provides the technology needed for future generations of computing systems. ARPA's larger cost shared projects with industry have accelerated the delivery of new generations of commercially available machines. Commercial systems can now execute codes at 100 Gops and scale to approximately 300 Gops (billion operations per second). Our strategic goal is to bring about a unification across the prevailing models of computations within the scalable computing framework. For example, Cray's new T3D system, introduced last summer, went from concept to operation in only 26 months, and represents a new approach for integrating vector and massively parallel processing technologies. Machines of this class from Intel and others are now reporting results of significant economic and societal impacts, such as analyzing seismic data to uncover new oil fields in the Gulf of Mexico, simulating large molecular interactions leading to a better understanding of the enzyme-based cause for Alzheimer's disease, reducing design time in the automobile and aerospace industry, yielding major cost savings in the financial industry, and many other results my colleagues, whose Agencies are direct users of this R&D, are reporting today.

In addition to these larger projects, important research results have been consistently emerging from open research efforts. For example, prototype demonstrations at Stanford and MIT highlighted the design of a general purpose shared memory multiprocessor and approaches to compilers and systems software. These efforts have not only influenced commercial products at Thinking Machines and Intel, but also companies like Silicon Graphics. In the process of early evaluation, our joint efforts with NSF are driving new system software development at the Pittsburgh Supercomputer Center on the Cray T3D, and at the Cornell Theory Center on the IBM SP-1/2.

Microsystems Design Technology has developed important new design tools and is moving toward computational prototyping of systems. 3-D simulation and modeling is being used to design semiconductors and system components in areas that have never been possible. Synthesized communication subsystem components designed with advanced tools from this program have now found their way into a variety of commercial and military products. Caltech technology research, after a ten year long focus on high performance interconnect technology as the basis for multicomputing, has now been spun out in a new startup called Myricom, specifically to commercialize the technology.

Scalable Software has been a critical component of the program, and the efforts established late in FY91 are now starting to show results. The same operating systems concepts demonstrated under ARPA sponsorship on very large machines, have been adopted by leading manufacturers of workstations as well. Microsoft's WindowsNT operating system was greatly influenced by the ARPA-developed Mach. And Mach is the OS running on the vast majority of today's massively parallel processors. New parallel programming environments for standard languages like C++ and

Fortran, as well as new experimental programming languages, have been demonstrated and are needed to stimulate new applications across various architectures. The advanced compiler technology developed under ARPA sponsorship at Rice has now been adopted by Convex for its latest generation of scalable systems. Finally, libraries and their supporting technology for scalable systems have led to wide area file systems and prototypes of approaches for architecture independence.

Networking has been a dominant theme in the program, from the earliest days of packet switched ARPANET to today's Internet with millions of users. ARPA's program, in conjunction with the Gigabit Testbed research with NSF, has demonstrated medical, terrain visualization, and modeling applications on gigabit class networks and is demonstrating optical and ATM technology operating over fiber and satellite media. This has accelerated the convergence of ATM interoperability across the telecommunications providers and has led to the earlier offering to commercial users of high data rate services than originally expected. In addition, information security is an important dimension to the ARPA program, and is becoming increasingly important to the future of computing and communications. Today, efforts have focused on trusted operating systems, mechanisms for secure access to Internet services, and establishing the Computer Emergency Response Team (CERT), which I know this committee is familiar.

Early R&D results have been used to support and demonstrate a number of techniques for distributed digital libraries, which are rapidly becoming the basis of larger scale deployment across the Agencies. ARPA has worked closely with the Library of Congress to develop an experimental electronic copyright recordation system that is laying the foundation for intellectual property rights management in the digital age. The ability for applications, operating systems, and supporting environments to exploit these capabilities across geographically dispersed areas is the basis for implementing the NII. Furthermore, by working with industry, this work provides important insight into the technical approaches that will ultimately be adopted throughout the field.

STRATEGIC DIRECTIONS AND FY95 BUDGET REQUEST

The opportunities to leverage results from the HPCC program are unprecedented, and, in my opinion, is critical to maintaining our nation's strategic advantage. Virtually every other major program in government and industry is relying on computing and information technology, and the demands are increasing.

On the financial side, there are two complementary views of the budget - the Federal HPCC Program crosscut and the DoD Descriptive Summary requests. The detailed descriptions of each DoD project contributing to the Federal Program are contained in the recently distributed HPCC Implementation plan. ARPA's FY95 Federal HPCC request is $357.4M, an increase of $58.4M over FY94. I'll briefly describe the major strategic directions of the FY95 program, focusing on the key project, High Performance Computing (ST-19).

First , a major increase is planned to support the administration's National Information Infrastructure. The HPCC program's research investments in advanced computing and networking technologies lay the foundation for this critical national-scale enterprise. Under the HPCC program's new Information Infrastructure Technology and Applications (IITA) program component, we are developing a layered architecture for the National Information Infrastructure that will ensure high leverage through a common set of end-user and applications-developer services that bridge the network "bitways" and the National Challenge applications. This, of course, is being done in conjunction with industry, academia, and other agencies. Let me describe each of these layers, bitways, services, and applications, in more detail:

- The bitways are the "pipelines" or distribution system through which information will flow between repositories of information and user applications. We will continue to build upon our investments in high performance networking, moving from gigabits to tens of gigabits per second, for this technology will be needed to form the "trunk" lines of the information superhighway. In addition, we plan investments to create the "feeder" line technologies, including wireless network extensions and hybrid networking technologies that can integrate fiber and coax cable distribution bitways.

Last February, the Vice President visited a high school in Cupertino, CA whose Internet access was provided over the local cable TV system -- using an early prototype of hybrid networking technology developed by ARPA.

- Our greatest technology challenge, and the major new focus of the ARPA 1995 program, is to develop a robust services layer that provides critical building blocks for the National Challenge applications. Services are bitway-enabled technologies that provide widely available, network-aware, adaptive interfaces upon which to construct complex, distributed applications. These include, for example, ways for applications to negotiate for and adapt to the available quality of service they need or can pay for; a charging infrastructure for access to network resources; technology for privacy-enhanced, secure, and trusted processing on behalf of users and applications; and more protected ways to interface private bitways and their information resources to a common, shared "open" network like the Internet.

- The ultimate value of the NII will be determined by the quality and effectiveness of the applications delivered to the end-users, the American public. ARPA will focus on those dual-use National Challenge applications with a high degree of relevance to our Defense mission. Specifically, these include design and manufacturing for affordability, information technology support for improved trauma care and health care information processing, simulation-based education and training, and planning-based crisis management. Further, ARPA will develop the underlying technologies for digital libraries -- large-scale, distributed repositories of digital multimedia objects like text, maps, and imagery -- and electronic commerce -- the technology for electronic data interchange, automated sourcing and brokering, and interfaces to electronic payment systems -- that will play a critical role in enabling these other National Challenges.

Second, we plan to increase the software technology investment and, in collaboration with other Agency efforts, meet the critical need for technology to enable applications on a variety of platforms. A series of collaborative pilot projects will be established aimed at unifying compiler technology, run time systems, and support for scalable input/output. Software has always been an integral part of the ARPA HPCC program, and we currently estimate that approximately 55% of the FY94 program of the High Performance Computing project is software related. However, this work is needed to stimulate the development of new codes by Independent Software Vendors (ISV), as emphasis in the market begins to recognize this viable new technology.

Third, we plan to increase investments in critical areas of scalable systems. An important variant of scalable systems, called embeddable systems, will receive increased funding in FY95. Embeddable systems are those that are packaged to withstand harsh environments that may be found in typical military applications, such as on-board unmanned air vehicles (UAVs). Embeddable systems will use the same software as used in the commercial variants, thus reducing the cost of developing these systems. In addition, the early evaluation component will be reinstated, to insure the earliest possible feedback between the industry and users of new systems. While continuing to push high performance computing development for scientific and defense applications, alternate scalable configurations will include high performance servers, clusters of workstations, or networks with many vendor's computers. We plan to add important new capabilities to families of computing systems such as real time response, and scalable input/output to meet the needs of the systems users.

Fourth, the networking world must be prepared for a balanced environment of information security, high performance, and advanced services. We plan to increase investments in network scaling issues, information security, and couple these efforts to our IITA component.

Fifth, we plan to apply HPCC to design technology and its experimental applications to computing. As computing systems become increasingly complex, there is a cost, time, and scientific benefit to prototyping the system in software, rather than only use traditional experimental prototyping methods.

Finally, I'd like to point out that supporting detail is provided in the HPCC Implementation plan, as well as DoD's Descriptive Summaries, particularly ST-19, High Performance Computing.

PROGRAM ASSESSMENT

ARPA reviews its HPCC efforts at many different levels to insure consistent evaluation in a dynamic R&D environment. Projects produce reports on a regular basis that are reviewed by program managers. An annual process of updating accomplishments, milestones and project plans is tied to the funding process. Site visits, project meetings, principle investigator meetings, and regular interactions with industry, industrial groups, academia, and other government agencies occur frequently. Office directors and program managers develop plans and milestones that are approved by senior management during the planning and budget cycle. New programs and ideas are proposed during this process. In conjunction with yearly funding decisions by ARPA and DoD, senior ARPA technical management critically reviews program areas, plans and accomplishments. Guidance is provided to reflect programmatic, technical and funding directions.

The net result is reflected in Agency plans, Descriptive Summaries, and other products produced in the program such as the Implementation Plan, which reflects the consolidation of overall FY95 program execution goals and milestones.

SUMMARY

I strongly believe that the Federal HPCC program represents a unique and unprecedented effort across the Federal government. It has been very successful in exceeding its goals, and ARPA is proud to be a key Agency in this important program. As we look to the future, the investments we make today and in FY95 will provide the important foundation for an advanced information infrastructure in the 21st Century. Thank you.

Mr. BOUCHER. Thank you very much, Mr. Toole, and we express our appreciation to each of the witnesses for their well prepared and well presented statements here this morning.

One of the criticisms that has been raised by the private sector relates to the lack of opportunity for nongovernmental advice, comments, input into the planning process for the HPCC program. In order to formalize the receipt of that kind of advice and recommendations, the 1991 Act required the appointment by the President of an outside advisory committee. To date, that committee has not been established despite the urgings by this Subcommittee to the contrary.

My question to you really is in two parts. First of all, just comment, if you would, about the opportunities that are available for nongovernmental participants, those who are interested in the program, to offer advice and suggestions. Is there some process of a formal nature that you have established, perhaps through your office, Dr. Lindberg, but others can comment on this as well, in order to receive that advice, set of advice and recommendations? And do you think that the absence of the formal external advisory committee is a serious impediment in terms of linking your activities directly to private sector users and getting their advice in the planning process?

So it is a fairly broad question about your ability to have meaningful interchange with those external to the government.

Dr. Lindberg.

Dr. LINDBERG. Mr. Chairman, I think that it is a serious impediment not to have an advisory council, and I have said that repeatedly. I am assured as of last night that the White House is in the final stages of appointing an advisory structure. They are going to use the mechanism of the National Science and Technology Council, this new mechanism, you know, that replaced HPCC, and that has an advisory structure which I guess is appointed by the President's Science Advisor, Jack Gibbons. They have settled on a list, and they are contacting the persons to find their availability. So then a subset of those would be advisors to us, and I think that would satisfy the fact of law and would be a tremendous help to us.

In the interim, as I tried to indicate, we have scheduled a string of meetings that have been very, very useful to us, and we will continue to do that in any event. But in the ideal circumstance, those would be overseen by the group of outside advisors who are legally proper to do that, and they could have regional hearings as well.

Mr. BOUCHER. Dr. Lindberg, how do you conduct these meetings presently? Are they organized by your office, or are they organized by the individual agencies that participate in the HPCC program?

Dr. LINDBERG. Well, we do it jointly. They are organized by our office, and I sign the invitations, but we all agree on the constituents. We have, of course, tried to draw on industry associations where we could. CSPP is one of those obvious ones.

In the case of the telecommunications industry, it is a very, very complicated industry, as you probably know better than I, and in that particular case, we got advice from Bob Kahn on sort of the structure and how we could get representatives from each of the right components.

Mr. BOUCHER. Do you get any complaints that you are not inviting the right people?

Dr. LINDBERG. We haven't so far. I guess there is a self-complaint because in the third group of software, I don't think we ever felt that we really understood the field suitably and we couldn't identify anybody like Kahn who would step forward and say that he or she understood it thoroughly, so we were not exactly shooting in the dark, but we definitely still need some advice. That would be one of the first things we would ask our advisory council about.

Mr. BOUCHER. Let me just get others to comment on this general process.

Do all of you have the sense that these interim measures to establish a means of getting advice from those external to the government has been satisfactory, and how would you recommend that it be improved if you could offer recommendations on that?

Dr. LINDBERG. If I could just add one thing, Mr. Chairman, there is a tremendous amount of contact with industry at the working level. I think others have indicated that. The consortia and the working level arrangements bring a day-to-day contact. But I was just addressing the more formal structure. We need both, of course.

Mr. BOUCHER. Right.

Dr. Nelson.

Dr. NELSON. I certainly agree with Dr. Lindberg, and I would say that perhaps the biggest problem with the current mode is that despite our best efforts it is somewhat random. We hear from the people that are at the table. Typically they don't write written reports. An Advisory Committee offers the discipline of a written, well commented, thoroughly reviewed report and, therefore, can represent more of a consensus in the field.

Mr. BOUCHER. All right.

Any others?

Mr. Toole.

Mr. TOOLE. Yes, Mr. Chairman, I would like to agree with Dr. Lindberg as well and add a couple of things.

Technology is moving very rapidly here. On the order of months, things are changing, and I think we have got to have both a very tightly coupled, informal process along with the formalities that we are really referring to here so that we can act and understand what is really happening out there in the industry.

While advisory groups themselves can get a very well thought out strategic view, I think there is a danger here if we are not careful, we will not be able to couple in real time well enough with the industry groups; and as a result, I think at the working levels that were referred to earlier, each of us and the program managers in each of our respective organizations are responsible to deal directly with industry groups, advisory panels, and such that other areas that are out there relate to our work and deal at the CEO and VP level on a day-to-day basis, and that has been an informal process but I think a very effective one that, coupled with the formalities of instituting the advisory panel, will give us the breadth and strategic direction that I think we all want.

Mr. BOUCHER. All right.

Yes, Dr. Ciment.

Dr. CIMENT. I agree with Dr. Lindberg that there is a need for a coordinated advisory committee, and we have supported that. In the absence of that, NSF has had a vigorous program of including advisory inputs from various sources.

There was a Branscomb Blue Ribbon Panel on High Performance Computing which looked at the entire HPCC program in the context of supercomputing primarily. That was reported to the National Science Board. Moreover, we have internal advisory committees which have a broad perspective given the nature of NSF's program, and in fact, oftentimes, our colleague agencies are asked to come in and talk about various aspects, and Dr. Lindberg has addressed the size of the advisory committee in particular.

Moreover, there is a Federal Networking Council Advisory Committee where the interagency efforts in networking are represented, and there we have considerable industry participation.

However, in summary, I still think that a concerted, organized effort with more discipline would benefit the entire effort.

Mr. BOUCHER. All right.

Yes, Mr. Holcomb.

Mr. HOLCOMB. Yes, I would like to comment also. From the perspective of our agency, we actually have an agency NASA Advisory Council that set up a specific formal task group to review our program, and we also use informal mechanisms. Particularly in our computational aeronautical sciences project, we have engaged the aerospace industry specifically in reviewing the program.

I would like to comment that I believe that the informal mechanism, review mechanism, is as important as the formal one because of the dynamic changing particularly of industry goals, industry needs, and the pace of the technology. I believe the program needs both.

Mr. BOUCHER. Isn't there a risk, though, Mr. Holcomb, that if you rely unduly on the informal mechanism, that you may tend to favor a few selected firms that have long standing ties with the agency and where the personalities involved may be particularly close, to the exclusion perhaps of other companies that don't enjoy that close relationship and have the opportunity to offer what may, in fact, be very valuable comments? Isn't there that risk?

Mr. HOLCOMB. Well, two—I think, first of all, I believe both mechanisms are important but, secondly, even in the informal mechanism we periodically go back to the entire industrial sector and query the whole sector periodically, and, yes, you do have in the intermediate period a smaller set of groups who are very vocal and very active, but we periodically, even with our informal groups, go back and try to contact the entire sector at one time.

Mr. BOUCHER. Well, thank you for that answer.

Let me ask this of I suppose Dr. Lindberg first but others as well.

Do you have any special means, apart from just the regular announcements of program solicitations, by which you let the private sector know and, for that matter, the university community know of the various projects that you are proposing to undertake so they can make requests or submit grants and the like? Do you have a special means of doing that?

Dr. LINDBERG. Amazingly enough, these computer networks and bulletin boards are very heavily used by universities. They have caught on to that quite quickly, and so we find that this Mosaic server with its reference to the agency servers is extremely heavily used, something like 14,000 times a month. That isn't sufficient. I mean we publish things in the Federal Register, and we give talks at meetings and so forth. But I would say that for the most part, the institutions that are able to participate effectively in these programs are pretty well informed through this.

Now at the lower level of kind of getting people hooked up, institutions that aren't part of Internet, the NSF has taken the lead in a connections program, and we have found at NIH that we are better off to transfer money to NSF and join their connections program for hooking up, let's say, hospital libraries. So we are working both sides of it.

Mr. BOUCHER. All right.

Any others care to comment on that?

Let me ask you this. Assess, if you would, the value that has been added to the various programs you have undertaken and the goals of your program by virtue of the fact that it is a coordinated interagency undertaking. What, in other words, has been the value that multi-agency participation has conferred that you would not have received just with the individual agencies pursuing their separate agendas?

Dr. Lindberg.

Dr. LINDBERG. If I could make a brief suggestion, I think that we would have ended up with nine or ten separate Internets if we hadn't joined forces in support of the lead that NSF took, and so I would like to ask Mel Ciment to follow it up.

Mr. BOUCHER. Dr. Ciment?

Dr. CIMENT. Well, I think we have spoken to the networking aspect already, and I would like to expand on—I think the Grand Challenge applications is an area where we talk about challenges that are well beyond the computational and algorithmic formulations of today. The various agencies working together have had significant progress in this area in terms of focusing critical mass of workers who are willing to work in new scalable parallel environments.

It turns out there's a lot of tools development and a lot of applications tools that are necessary in this, and these teams are formed across the agencies. This gives life to the various enterprises that need to support these activities, and that is why we have seen the growth of so many different Grand Challenge applications areas.

The list is growing—you know, Time Magazine published a list of 10 Grand Challenges, but of course what we meant by Grand Challenges is that anything that requires a solution to a fundamental science and engineering problem, and these are fundamental problems like in human joint mechanics. Disciplines that need to model aspects like that just a few years ago didn't have the—I would say the support to go ahead and undertake such enterprises. Today they find groups across the agencies that have joined together to participate in these things.

Mr. BOUCHER. Yes, Dr. Nelson.

Dr. NELSON. Let me add some other examples.

First in data storage, the National Storage Laboratory, which has the participation of several agencies, a lot of universities and laboratories, is struggling with how to deal with all the data that these Grand Challenges are producing, and I think without the cooperation of the several agencies we wouldn't have the critical mass that now has gotten first products out and beta tests.

Second is an interagency program being developed on scalable input/output. It is easier to compute than it is to move data around, as it turns out, and unless we can deal with how to move the data around both within a computer and among computers and to users, all the compute power we have may not work to the extent that we hope it will and that this scalable input/output program requires the perspectives and the funding of several agencies to be successful.

And finally, I would mention in connection with the Internet, the gigabit R&D program where several agencies are participating and the private sector also has brought again perspective and funding that has given us now some pathways toward the gigabit per second data rate, as Mel Ciment mentioned, where we are sitting in our national networks now at about 45 megabits per second and we are struggling to get to 622 megabits per second. But the data demands today really do add up to gigabits.

The Internet is seeing some constriction in band width, or in the effective band width, because of things like Mosaics, so we have got to stay ahead, and things like the gigabit R&D are the path to do that.

Mr. BOUCHER. All right.

Mr. Toole.

Mr. TOOLE. Yes, Mr. Chairman, I would just like to add a couple of things.

I think if you look over the past seven years you will notice a profound change in the way these agencies can actually work together. The first early days, I think my colleagues will attest to, were rather hectic in terms of trying to iron out responsibilities and directions. Today basically at a HPCCIT meeting, our executive meeting, or a telephone call with my colleagues, we can really come to grips and know precisely where we want to move in that direction.

The gigabit testbeds were a perfect example, but the speed by which we can plan across agencies with these people and the confidence that we have built up I think is really unprecedented. I am very proud of that. I think the collaborations with NSF, NASA, and DOE where ARPA is concerned, we have realized where our centricities really are, and our strengths and weaknesses, and as a result we have put together more effective programs and projects.

Thank you.

Mr. BOUCHER. All right.

Yes, Mr. Holcomb.

Mr. HOLCOMB. I would just like to add one other comment, which involves a couple of agencies that are not here today.

In particular, we have currently out on the street a solicitation for access to remote sensing data with broad public use. This solicitation involves both NOAA and EPA data sets, and I think with the opportunity to work with the other agencies we are able to pool our resources and in this particular area bring together the best satellite imagery products that the public sector has for use by the private sector.

Mr. BOUCHER. All right.

Well, I hear you saying that there is a synergy created by having these agencies work together which obviously would avoid duplication of products that might occur in the absence of coordination, but that also perhaps creates a greater whole than the sum of the parts in terms of the effort overall. Given those benefits, I wonder if you could make some estimate as to how long you think this project should remain a coordinated interagency initiative. Have you given any thought to that, Dr. Lindberg?

Dr. LINDBERG. I think the best answer is to say that that is essentially our top of the agenda for the next year, is to figure out what answer should be given. We are aware—we are all quite confident that the program will achieve the goals that it set out to achieve. So the question then remains, should this momentum that has built up be directed or given a yet further goal or merely disbanded. We won't need this structure to achieve what we set out to achieve because it will be achieved.

Now fitting it into the National Information Infrastructure is yet another question. Being sure that the whole country benefits from it is yet another dimension. Perhaps going beyond it to even higher technical goals is under consideration. And we want to work all that through and get as much advice as possible in order to bring to the committee a recommendation.

Mr. BOUCHER. It is fair to say, I suppose, that the goals of the program are evolving over time, are they not? I mean do you still see yourself today as having precisely the same set of goals that were enumerated in the High Performance Computing Act of 1991?

I guess I would answer my own question in part by saying obviously not, because we now have an applications program even for the current fiscal year and with a much larger recommended applications program for fiscal year 1995.

So given that evolution over time, would it be reasonable to assume that perhaps this coordinated initiative should continue for the foreseeable future, some time to come, perhaps make it permanent?

Dr. LINDBERG. We think—we are not yet prepared to make a recommendation, but we do agree wholeheartedly with you that the addition of the NII in the applications phase is a major, major change, and that can't be achieved quickly and it can't be achieve in the first five years.

I agree with you, the goals have changed. I meant that technically, the fast computer, the fast network, the functioning software probably will be achieved, although, as we are learning, the software is itself a pretty big impediment.

So I would imagine that the major undone work will be extending this to the whole country rather than just the top 1 or 2 percent whom we serve right now.

Mr. BOUCHER. All right.

Dr. Ciment.

Dr. CIMENT. Mr. Chairman, without commenting on the permanent nature of such an extension, I would say that one has to take note of the fact that this coordination process is with some cost. Each of the agencies, aside from developing budgets within its own agency context, has to do this interagency planning which is considerable, and from my perspective personally, I feel that is the value added that I gain from working with the agencies in the high performance computing program. For the near future, I would not want to give that up for the benefit I gain from being able to share our plans and develop common plans. So I would personally recommend that this be extended and reevaluated as the technology develops and as the programs develop.

Mr. BOUCHER. All right.

Do others care to comment?

Dr. Nelson.

Dr. NELSON. Just briefly, there is a culture as well as a specific action that we are talking about. The National Performance Review, I think, has referred to concepts such as we describe, the virtual agency, and I think the people who have participated in the HPCC program have caught that new culture. Maybe we did it five or ten years early, but I think as individuals it will be hard for us to go back. We may be doing different things. It may have different names. But we are proud of the fact that we are what the NPR was talking about, and maybe we go there first.

Mr. BOUCHER. Okay. Very good.

One of the questions that we get, and in fact it is sometimes asked by our colleagues in the Congress, is, what has this program accomplished? I mean here we are, I guess, three years into the program now. What have we achieved? So if you would, here's an opportunity for you to brag about your successes. Let's point to some of the specific accomplishments, and, to the extent that you have been able to transfer that progress into the private sector, this would be a good opportunity to underscore that as well.

I heard a reference, and I have forgotten who said it now, earlier during the testimony, about the development of asynchronous transfer mode and SONET technology for packet switching. Maybe you could elaborate on that to some extent and talk about the extent to which that has been developed or perfected as a consequence of the HPCC program, and mention, if you would, other notable successes that we can now point to as a consequence of your work.

Dr. Nelson.

Dr. NELSON. I will probably do my answer in different pieces. I will pick up with what you mentioned, ATM and SONET, and then I will yield to my colleagues.

The DOE NASA procurement of high speed ATM and SONET services has been referred by outsiders, by participants in the industry, not by us, as a landmark by making ATM respectable for operational long-distance networks, and the vendors who have bid on that—as I think you know, that contract is still not let—have used the respectability that our solicitation, by sophisticated people again—and that was part of the importance of the solicitation—to market successfully ATM services to their private sector customers.

Mr. BOUCHER. All right.

Mr. Holcomb.

Mr. HOLCOMB. I guess I would like to comment on another area of where I feel very strongly that the HPCC program impacted industry.

Several of us met back in about, it was the summer of 1990, with John Armstrong from IBM at that time—he was the CEO and President of IBM—to discuss their supercomputing strategy. The position of IBM at that time was that all their eggs were in the basket of Supercomputing Solutions, Incorporated, a small company that was founded by Steve Chen, and the target was to develop a conventional vector supercomputer, and the company was financially backed very strongly by IBM.

In February of 1993, about two years into the program, we again met in a closed session, this time with Dr. Irving Wladlawsky-Berger, who is the Director of IBM's Supercomputer Division, and he indicated at that time, because of the HPCC investment in scalable parallel computing, IBM made a decision to invest heavily and enter the scalable parallel computing field, and I believe that those events have led to the new product, the new IBM SP-2, as a very, very capable general purpose parallel supercomputer.

So I think this is an example where we have made a major company change its mind technologically and move into an area of importance to the program and I think move the country ahead.

Mr. BOUCHER. Okay.

Dr. Ciment.

Dr. CIMENT. I would like to expand on Dr. Holcomb's remarks. In fact, in the case of IBM I would say that the Cornell Theory Center under the direction of Dr. Mal Kalos, has been a significant influence in developing IBM strategy in that particular area, and in fact one of the first scalable systems, IBM's entry into this arena, was introduced at the Cornell Theory Center last summer, and I think this speaks to the larger development that has happened here actually.

When this program started—actually before this program started, when our supercomputer centers were just being recommended by the Lax report over a decade ago, there were many people who thought that these centers would not really work in the academic environment. There were those who said you should just get your cycles from job shops and the academic world would never really get involved. I think the greatest success of the program has been the change of culture within academia, working with the National Labs, working with industry.

This cultural change has happened over many years, and so people haven't really noticed it. But if you think back to what the centers were like, you know, scientific—the computing centers, rather—academic campuses, a decade ago, they weren't involved with vendors, they weren't the people where new systems were introduced to sort of shake them down, those were done at the agency laboratories, frankly, and now the academic world has entered that arena also, working competitively, cooperatively, and has become a major player in this whole environment, and I think that has been the major transformation.

Mr. BOUCHER. Mr. Toole.

Mr. TOOLE. Yes, Mr. Chairman. The story that Lee Holcomb just related I think has occurred over and over and over again in basically many, many different companies from Cray Research, Intel, MasPar, Kendall Square Research, and in others in terms of, you know, assisting and talking to them about how the actual evolution of scalable parallel systems really is evolving.

I think the point to be made is, five or seven years ago it was very unpopular to think that gee, while we have had some university research examples, it really was not going to take off, but probably between three and four years ago just before the genesis of the HPCC federal program, there was almost uniform consensus across the major industrial players that scalable high performance computing is really going to be important to them strategically now and in the future, and that is the result that we are starting to see.

If we add to that the actual research efforts that have been coupled with industry; I think it is very important to give you some examples. Cal Tech, for example, had done a lot of work funded by ARPA and other agencies having to do with high speed interconnect technology that was important. That ultimately became part of the basis for the Intel multicomputer whole system of operation over many, many years, coupled with the laboratory systems.

In addition, we were seeing now couplings with Intel and Stanford as a couple of examples. Another example, Cray Research has been involved very much in many of the technical discussions and meetings and have used the results of performance analysis studies that are coming out from many of the university bases, from MIT, that has been very important to their particular future plans.

From the small company side, there are also spin-outs that have been made really relative to much of the technology that we have invested in. Tera Computer is a computer of a shared memory processor that is basically going to be coming out as a small start-up. Miracom is yet another small company that Chuck Seitz spun out from Cal Tech and is commercializing some of the technology.

So it is not just the large systems, but some of the small systems and the small start-ups, I think, that have been stimulated by this.

Thank you.

Mr. BOUCHER. Dr. Nelson.

Dr. NELSON. Yes, I said I would want to get back with more of an answer.

Let me talk about the value of computer science and its immediate application in industry, in this case the automotive industry, and I will do it with two examples.

The first example is work between Sandia Laboratory and Ford Motor Company. The computer science advance is better ways of generating what is called computer meshes. Since computers do things by point, you have to lay out a computational mesh that covers your problem in order to do the calculation. It is not very exciting, but most of the time taken in crash test simulation has been doing the computation mesh that describes the car.

Now the computer science advance called paving has reduced the time to do those meshes from six to eight weeks per case down to so far a few days, and we think it will be a few hours as it becomes automatic.

The value of crash test simulation is both in dollars—the cost of a single crash test if you do it physically with a real car is $50,000 to $500,000; the cost to do it on a computer is a very few thousand dollars. More importantly is the time. I mentioned it was six to eight weeks to do a computational mesh, but it can be six to eight months to do a fully instrumented car crash test, and so the change from many months to do a crash test in the real world to a few weeks or even a few days to do it in the computer allows time to market.

So the point is, a rather specialized computer science advance gives Ford Motor Company and other U.S. manufacturers who use this cost saving and time to market advantage.

The second example that I wanted to give is something called automatic differentiation, which is a very technical piece of computer science. Basically, it is a way of optimization. It is a way of getting your code and your answers optimum.

Now the Argonne National Laboratory, working with General Motors, has applied Argonne's computer science advance called ADIFORS to GM's stamping of door panels and other sheet metal, and the value of using this technique is that GM is now able to make more accurate stampings that are stronger because they don't stress the metal and are more suited for things like internal stiffeners.

So again, this computer science advance, automatic differentiation which would be in arcane journals is, in fact, an important tool for General Motors to make better, safer, cheaper cars.

Mr. BOUCHER. And, needless to say, it would be unrealistic to expect the automotive industry to come up with these advances on its own.

Dr. NELSON. To my knowledge, the automotive industry does not invest in computer sciences.

Mr. BOUCHER. Exactly.

Would it be reasonable, however, to expect some of our computer companies to come up with those advances on their own?

Dr. NELSON. Now I am more speculative, but to my knowledge computer companies have a minimal investment in computer science. That is something that is more carried out in universities and in National Laboratories. So while not impossible for a computer company to come up with it, my experience has been it is seldom.

Mr. BOUCHER. All right.

Dr. Lindberg.

Dr. LINDBERG. Maybe I could just make a very tiny addition following this line of thought.

The division of work is such that even when a company decides that computational work is very much in their interest, they would prefer to collaborate.

The Lilly Drug Company worked with NCSA, the National Center for Supercomputer Applications, to look into and investigate initially in a rather casual way the use of high performance computers and ended up concluding that it was very important to them or could be very important to them in drug design to model a molecule, the three-dimensional structure, the receptor, and the molecule.

It was assumed by the center that when they made a decision to put in their own supercomputer, that was kind of the last they would see of them, they were converts to the cause, and that was the end of it. But quite the reverse actually happened. The pharmaceutical company continued to pay their dues and continued to associate with and meet regularly with the supercomputer center because that is where the ideas are.

The reason to have their own machine was partly because of the utility and to have it available, partly because of the security involved in doing their own experiments, but the ideas are going to continue to arise in the universities.

Mr. BOUCHER. Dr. Ciment.

Dr. CIMENT. To follow up on your question about whether the companies—the computer industry will do the research. In the Branscomb panel which I referred to earlier, there was analysis done of where these new architectures have come from, and in many cases it was pointed out that the basic architectures and compilers and structures were seeded in the research efforts of over a decade ago, research that had been funded by ARPA, NSF, and other agencies.

Mr. BOUCHER. You have talked a good bit in answering this question about the advances made in the computational sciences and in computer technology. The other aspect of the HPCC program is high-speed networking. A reference was made to the ATM and SONET packet switching advances, but I would like perhaps, Dr. Lindberg, to start with you, if you could explain to some extent the work that is done jointly in the testbeds with the telecommunications industry, telephone companies in particular, and any advances that are being made on that front.

I think you mentioned that the telephone industry, taken as a whole, is investing more in this than the Government. So what is the value of the Government's investment? Has it leveraged that private sector investment? Would the private sector investment be occurring in any event? And what kind of advances in fact are being made through the collaborative effort?

Dr. LINDBERG. I think it is a very interesting situation. To some extent we are merely speeding things up. I think to another extent things are happening that wouldn't happen without the program.

But you will recall that one of the purposes of the program was to play towards America's preeminence in this field and to retain that. Knowing that of course, advances will occur in computing and communications regardless of this program, but they won't necessarily occur in the U.S. I am quite convinced that that is a real phenomenon.

When this program started, for example, conceptually in 1987 and 1989, the papers were full of reports that we had lost the computing industry to the Japanese and the fastest chips were NEC and so forth, and that isn't true any longer. Al Massey reported at a hearing, I think in this room, in the last year that in the previous year 89 percent of all the high performance computers in the world sold were American, and certainly in communications we are doing extremely well.

In the Aurora testbed, just to take a specific case, that involves Pittsburgh and Philadelphia and Boston—MIT—There are three phone companies, as you would imagine—MCI, Bell Atlantic, and Nynex. All of these are committed to SONET/ATM. I think that did derive directly out of this program. But that doesn't mean that they are interoperable. In fact, the SONET switches had never met from the two companies, MCI and Nynex, sort of exchanging these billions of bits per second in midair as it were, until it was done at Aurora Test Site in Philadelphia. So both of those companies told me that that was a remarkable advantage to have a real job to do. These are computational jobs, fortunately, not lives depending on them.

Yet, on the other hand, it is possible for the switches to fail, it is possible to do the engineering experiments under this controlled test environment. It was extremely valuable to them. IBM is also a part of that network, as is University of Pennsylvania, University of Arizona, and MIT. So I have seen that particular thing happen.

In another testbed site, the Vista site in North Carolina, there General Tel and Bell South are the regional phone companies, as it were. Now both of them have their own technology and strategy and so forth, but in actual fact the first time they had ever switched two billion bits per second is in a little box in Durham, North Carolina. It is actually midway between Durham and Chapel Hill. But that is where it actually happens in kind of a locked up building with everybody standing around somewhat in amazement, and everything else is very, very slick and professional. That particular box, you know, kind of looks like it did come out of an engineering laboratory.

So I think—in this really advanced area, I think everybody, including the communication giants, are feeling their way and definitely feeling the benefit of a testbed situation.

Mr. BOUCHER. And what the HPCC program has provided is the testbed environment in which these companies can collaborate. Is that right?

Dr. LINDBERG. That is exactly right. Those are set up more or less over a three-year period. I guess you would call it pre-competitive technology because in the end it is fully expected that the companies will walk away with marketable products. The world, of course, will learn the technology. But it isn't meant to go on forever and doesn't need to.

Mr. BOUCHER. All right.

Let's talk a little bit about general funding levels and the allocation of funds among the various agencies for their work. To take the current fiscal year, the high performance computing systems component is 18 percent of program funding; advanced software technology is 38 percent; networking technologies and network access for the R&E communities, 15 percent; basic research and human resources, 12 percent; and applications currently receive 17 percent.

I will ask in a few minutes the basis on which some changes have been made over time in these allocations, but before I get to that let me ask you if you generally agree that these allocations are proper, and do they reflect an appropriate balance in terms of the priorities of the program and the work that needs to be undertaken?

Dr. Lindberg.

Dr. LINDBERG. I guess I would have to say we think they represent a proper balance.

Mr. BOUCHER. I am shocked.

Would the other members of the panel like to comment on the appropriateness of this balance?

Dr. LINDBERG. I think you will notice that a fair amount of emphasis is placed both on software and on applications, and I think we have all indicated that we really do acknowledge that software is an eternal big problem, and we are told again and again by the commercial people that that is still the case, and we are also enthusiastic about really this committee's legislation in terms of the information infrastructure and the applications and are trying our best to get behind that.

Mr. BOUCHER. Thank you.

Dr. Nelson.

Dr. NELSON. I will start by saying amen to what Dr. Lindberg said and speak just for a minute to the ASTA component—Advanced Software Technology and Algorithms.

We do feel strongly the need for this program to be useful, and the utility, to a considerable extent, is expressed by people using the software, using the computers to do useful things. We have also found that this particular area is a very difficult one.

So the fact that the ASTA component has the largest amount of money we think is highly appropriate both given the emphasis on doing something useful and the need to confront some problems that have arisen in this area.

Mr. BOUCHER. All right.

Mr. Holcomb.

Mr. HOLCOMB. I will just second that same comment. Among the program, one of the roles that NASA plays, in fact, is to coordinate the ASTA element, and we felt right from the beginning of this program that the applications software technology element is the most important glue for the program.

We have invested from the beginning of the program about two-thirds of our agency budget in this area, and that has remained pretty much solid throughout the first four years of the program including our 1995 request.

But in addition to that, we have added, as have other agencies, this new IITA element, and that has come in, and its has largely been, in our case, at the expense of both the computing and the networking area which we think are progressing reasonably well.

So we are continuing to invest in that area as are the other agencies, and I think that this is where the Federal investment can have its biggest payoff.

Mr. BOUCHER. Okay.

Dr. Ciment.

Dr. CIMENT. Well, in terms of the NSF budget, the largest change has been the emphasis on IITA, and this has come at a time when there is a real opportunity to link the applications and the information intensive area to the computational skills and resources that have been acquired in the early part of the program. This creates a bit of tension between requirements on both ends, but I think it is a healthy tension that keeps the activity focused on essentials, and, for example, this year in 1994 NSF postponed a Grand Challenge applications group competition in order to be more actively involved in the IITA, but that will be reflected in more emphasis this year on National Challenges, as I said, and so

I think the balance that we have here creates the tension that is healthy for the program, and I think it challenges us.

Mr. BOUCHER. Yes, Mr. Toole.

Mr. TOOLE. Mr. Chairman, I also would like to concur with the previous speakers on this and like to add that I think the balance that you are seeing is really a dynamic balance which we have all put together across the Federal programs and coordinated with our respective agency heads.

I also believe and would like to add that the IITA component also is not strictly applications, it also is allowing us to put together some soft critical services that are needed to embody and enable many of these applications to occur. I think the testbed operations that the Federal Government can provide along that line, to enable some of these information infrastructure activities, are really a major challenge. ARPA's activity, for example, is really focused at that interface between those services and the applications layers and to work with the application developers to be able to make it robust and easy to develop those.

Mr. BOUCHER. All right.

Over time we have seen a change in the relative size of the major program components, and it has been in a state of flux over the last three years. The basic research and human resources component during that period has declined from 24 percent of the program in fiscal year 1992 to 12 percent of the program in fiscal year 1994, and the computing systems component has declined from 27 percent to 18 percent.

I have heard a number of references this morning to the need for more investment in applications software. I would assume that has been one of the factors that has occasioned this reallocation. But if there are others, I would like to hear that from you and get a sense from you as to what factors generally have been taken into consideration as these allocations in funding have changed.

Dr. Lindberg?

Dr. LINDBERG. Well, I think as John Toole suggested, these final numbers are the result of many, many—the sum of many, many small decisions. But perhaps he would want to comment on the hardware and systems side. On the third component, the basic research, that also is in no small measure ARPA, so perhaps he ought to field the question.

As far as the training goes, I think you are aware that for the—once again we are without the Department of Education, so education would be increased were there to be funds from education.

Mr. BOUCHER. Okay.

Mr. Toole, would you like to comment?

Mr. TOOLE. In terms of the overall percentage increases or decreases, I think you are also witnessing more agencies participating and expansion in terms of the pervasive computing communications, and I think in general the respective agencies in other programs within the Government are kind of perking their ears up to the HPCC program. These are the people that we also want to be able to work with and transfer our technology to, but without consuming them as part of the program, so we don't get too large ourselves.

In terms of the high performance computing systems components, I think the request is a very balanced one. ARPA did have a minor budget cut last year in the area where we had to regroup on a couple of different areas, particularly the early evaluation component and a couple of other ones. But the 1995 request really represents, in my opinion, I think across our agencies, sufficient activity to really go and to support both the new components that we are looking at across the Federal program as well as continue the innovation that I think is absolutely so essential and that we are seeing the results of in networking and computing that we have done for the past five years.

Mr. BOUCHER. The applications component represents almost 60 percent of the increase in the program growth for fiscal year 1995. Can you talk a little bit about the new applications research that you will undertake with that increased amount of funding.

Dr. Lindberg.

Dr. LINDBERG. Perhaps I ought to answer for a couple of the agencies that aren't here and then leave the bulk of it to my colleagues. But you will notice that NOAA has got a relatively high percentage of increase, although it is from a way too small number of dollars to a somewhat more adequate number, so in absolute numbers not very great.

They have done extraordinarily good work in application of this technology to the prediction of big storms, hurricanes in particular, and we have put those examples in the publications on a couple of occasions, but the Hurricane Andrew in 1992 they predicted extraordinarily accurately, but only to discover that they could have predicted it even more accurately if their big model had run it, the one at Princeton, had been able to run in real time rather than a day late.

Mr. BOUCHER. This raises another interesting question which I am going to divert to ask just at this moment, and I will ask you to return to your answer in a moment.

We basically see an intersection perhaps here between two FCCSET crosscutting multi-agency initiatives, that on global climate change and that on high performance computing, but to which of those programs do you assign the work you have just described? Is that an HPCC undertaking, or is that a climate change undertaking?

Dr. LINDBERG. It is HPCC. Actually, the person in charge of that from our point of view, Tom Pyke, is very well aware of the overlap with global climate, and he is part of that as well, so I mean we look to him to be our liaison with the global climate change.

Mr. BOUCHER. Okay. Sorry to interrupt you. Go ahead.

Dr. LINDBERG. No. That is a very good observation. I should have mentioned that.

In any case, NOAA then took advantage of their perhaps too small but still significant participation with HPCC to essentially get themselves better connected to utilize the computers that the other agencies were able to provide to them, and then next year, 1993, another hurricane season came along and another opportunity for the weatherman to shine, and in this case, Emily is the name of that particular challenge, and it did a thing that—the computer model there did a thing that was very, very spectacular.

Instead of just following the curve, so to speak, of the hurricane approaching the coast and extrapolating, which anybody at home can do with a protractor, this computer model said no, this hurricane is going to—said 72 hours ahead of time the hurricane will change course, turn at right angles, go out to sea, and not strike the U.S. at all, a very dangerous kind of prediction, and they weren't absolutely sure they could go with the computer model, but pretty much did, and it was correct. So their strategy of getting better computing was vindicated.

Now the increases that they are proposing here are to do two things. I mean it is really divided in half. One is to put more connectivity in NOAA so that all of their centers can connect to and utilize the Internet, because they really can't now, and the other half is to put in a heavy duty parallel processing computer at the Princeton Laboratory so that it will be computing in a timely fashion. So that really accounts for NOAA.

Mr. BOUCHER. And you define that as part of an applications program. It sounds to me like it is just the installation of hardware. Maybe I am missing something.

Dr. LINDBERG. Well, except that it is producing. We consider it to be a Grand Challenge, actually. Initially to predict weather 72 hours in advance of the weather is very much of a Grand Challenge.

Mr. BOUCHER. Well, the overall purpose certainly is, but it sounds like the particular components you have described for fiscal year 1995 fall more into the category of the addition of hardware rather than research with regard to a general application. Maybe it is a matter of terminology.

Dr. LINDBERG. Well, perhaps I should go a little further. They are essentially trying to connect all of that scientific laboratory really, what that agency constitutes, and to gather together the strength to utilize this advanced technology to do a number of things. Well, the first one is the 72-hour weather prediction, big weather, but there's other challenges ahead, and they have to work towards those.

Hurricanes are big storms, tornadoes are small storms. Tornadoes are what matter in the Midwest, they don't care about hurricanes, and in order to do that work in the Midwest, you need a much, much finer grid of observations, which means you need much more connectivity for the observations as well as the models to do the heavy computing.

Now beyond that—I mean I think these are applications, in other words—beyond that is a phenomenon that all of us know as wind shear, and that is enough to scare any traveler, that is even a finer grid, and those investigations are going on, in fact, out at Stapleton Airport in Denver by NCAR which is also funded by three of the agencies at this table.

So in order for a scientific agency like NOAA to progress toward the applications, they need connectivity, and they need access to heavy duty computing, and that is what they are trying to do.

Mr. BOUCHER. Okay. Thank you.

Other comments?

Mr. Toole.

Mr. TOOLE. Yes, Mr. Chairman. I described earlier kind of our concept of the bitways, services and the applications, okay, from the IITA perspective. Really, we plan investments in the bitway area, you know, feeder line technology, including wireless network extensions and hybrid networking technologies that can really integrate some of these things from a research point of view.

In addition, we really will focus on a number of the dual use National Challenge applications that we have described that have a high degree of relevance to defense; specifically, the design and manufacturing for affordability, I think we have got some insight to that; information technology support for improved trauma care; and health care information processing; simulation-based education and training and planning-based crisis management which we believe have a lot of dual-use applications.

The underlying technologies for digital libraries, for which we have joint efforts across our agencies, is also a very important part of this because that is an enabling technology. So in a sense what we are attempting to do is look for the lowest common denominator, if you will, and stimulate research and development that will have a much broader pervasive impact, and digital libraries I believe is one of those.

So these large scale distributed repositories, digital multimedia objects, and getting some early insight and basically fast forward into the future to see what that can be another generation or two perhaps ahead and get the industry involved in it we really believe can be an impetus to develop where the application interfaces are in this information infrastructure that we plan to develop.

Mr. BOUCHER. Mr. Holcomb.

Mr. HOLCOMB. Mr. Chairman, I think I would like to point out a few things I have mentioned a little earlier.

One area we are pursuing is the area of application of remote sensing data, and we have instituted a program to take remote sensing data and place that on to the Internet. It is a collaborative activity between NOAA, EPA, and the Department of Interior. Our objective is to put that out on the Internet on a basically marginal cost basis or free if you can retrieve it over the Internet.

This is—in response to your earlier question, this is in addition to the Earth Observing System, for example, which is providing data primarily to scientists who are doing the modeling activity, and this activity is aimed at the broad public, trying to serve other public use of the imagery and the satellite data.

To get started in this particular area, for instance, at the moment we are putting out GOES, GMS from Japan, and the Mediosat data from France distributed hourly over the Internet. So this is one particular activity that we are engaging in.

Our solicitation had more responses—we had over 470 letters of intent—than we had on the Earth observing program for science. So this shows what the broad public interest might be and the kind of data sets the Government can make available to the public.

Mr. BOUCHER. All right.

Yes, Dr. Ciment.

Dr. CIMENT. Just to pick one example of applications is the digital libraries initiative, if I can respond in my remarks about that. Here we are dealing with multimedia knowledge archive and retrieval systems, that we are asking researchers to think about developing tools interfaces, and the players in this—I am not at liberty to announce, you know, the awards yet, but we are in the process of selecting. Out of the 73 proposals received, we expect to make six awards, as I mentioned, jointly with ARPA and NASA, and the proposals have come in from the leading computer and telecommunications manufacturers, publishers, information processing companies, service organizations, and universities. The applications are as diverse as you can imagine from a variety of corporations, and in the second year, fiscal year 1995, we expect to expand the combinations of these awards with 25 supplements for additional parties to get involved with these initial awards.

Mr. BOUCHER. Okay.

Dr. Nelson.

Dr. NELSON. I would like to mention just one other application in the IITA area that the Department of Energy has just begun working on, but we are quite excited about the possibilities. This is the area of energy demand and supply management.

Now the fact is that today we use, on the average, only about one-half of the electric power generation capacity in this country. So on the average, about half of that goes to waste, and yet, as we experience both in the winter and in the summer, there are peak periods where we are using every bit and straining to keep up. So the opportunity is to smooth out both the day/night variations and, to some extent, lesser opportunity, the seasonal variations by using information and by using market forces, and this is, in a nutshell, what energy demand management is all about, it is using the information infrastructure to do time-of-day pricing, to give better customer information, and perhaps in the future even to allow the customer to pick from among the energy suppliers just as with the deregulated telephone industry the customer can pick from among long distance providers and perhaps in the near future from among local loop providers. Those are, of course, policy questions that would have to be looked at as one goes along.

Now in order to get started on this, the Department of Energy is cosponsoring with the Electric Power Research Institute a study that is being conducted now looking at the market potential. I really gave you facts and prospects which are those of proponents, and we believe the first step is to do an unbiased study to see what the options are. But as an example of future applications of this, we are, of course, working toward more introduction of electric vehicles.

Now, if you can imagine, when do you recharge the car? If you recharge it at five o'clock when you get home from the office, that is the wrong answer because that is one of the peak periods. If you recharge it at midnight, that is the right answer, and energy demand management can offer an economic, a marketplace incentive, not a requirement but a marketplace incentive, for the homeowner to have a timing device such that it recharges at midnight.

Now if the homeowner needs to go out that evening, it isn't, I'll say, fair to preclude the use of that car if it needs a quick charge at 5:00, but one needs to understand the financial implications of charging at 5:00 versus at midnight. So that is a specific example where the future introduction of energy demand management could avoid the construction of massive amounts of new generation and also of new transmission.

Mr. BOUCHER. And to facilitate the exchange of that information, you will be expending funds under the HPCC program for research. Is that correct?

Dr. NELSON. The only presently targeted expenditure is for this joint study with the Electric Power Research Institute. If the answer to that is favorable, as we expect it will, and also guiding with regard to areas of appropriate Federal investment, then, yes, we would expect to start spending some funds both in research and in data collection on private sector experiments that are going on now.

Mr. BOUCHER. Well, thank you. That is an intriguing area.

Let me ask a little bit about something Dr. Lindberg raised earlier, and that is the means by which we extend the benefits of the HPCC program to a broader population. I think you indicated that something like 2 percent—and I am not quite sure what that is a measure of, whether that is the private sector or the population as a whole—are participating now, perhaps deriving direct benefits.

The general question to the panel is, how do we go about expanding the benefits to a larger population, and, specifically, to what extent are the recommendations of the administration's NII task force, and particularly its Subcommittee on Applications, taken into consideration as a part of your planning process? Do you coordinate with the NII task force? Does its applications subcommittee make recommendations to you? Do you take those applications into account? And what other means by which do you intend to expand the benefits and the product of your research to a larger community?

Dr. Lindberg.

Dr. LINDBERG. The 2 percent I referenced was merely an estimate of the number of people in the U.S. population who use Internet right now. It seems to us when we are in a room of scientists that everybody is on Internet, but actually everybody isn't. So I think the concept of the NII that this committee has and the administration has I think is properly to extend the benefits to everyone.

We meet regularly. The same people at the table in our group are represented on these information infrastructure task force subcommittees. We also exchange invitations, of course, periodically with Dr. Prabhakar and Duane Adams to come and meet with us. So we are on a reasonably good contact.

The distinction between the two activities, however, is that the IITF is meant to design policy solutions which will enhance the speed with which the NII extends through the whole population, whereas we are charged basically to do the R&D. So we always try to keep that in mind.

Now since resources are limited, we want to focus on those things which will do the most good the fastest. But I think the quick answer is that we are well aware of their meetings, we participate and exchange electronic minutes and physical presence as much as we are able, and we pay a great deal of attention to that.

But the prime focus in IITF is to find policies which, if changed—Government policies and regulation—which, if changed, would help everybody to induce more private expenditure, more innovation, and more dissemination of the technology, so we endorse that view as well.

Mr. BOUCHER. All right. So you participate with the NII task force.

Dr. LINDBERG. Yes, sir.

Mr. BOUCHER. Tell me, if you would, Dr. Lindberg, what relationship you have presently or in the future intend to have with the National Information Infrastructure Assistance Program—I think I got that right—that is administered by NTIA. Basically they intend to provide fiber-optic linkages for educational purposes. They perhaps will attempt some other things. What tie do you see between the HPCC program and that effort?

Dr. LINDBERG. Well, I think it should get to be ever more explicit, to be truthful, and we anticipate that. We have again extended invitations to NTIA to meet with us, and we have, I guess, had four meetings.

But our understanding is that what they are doing is utilizing the funds, if appropriated, to buy off-the-shelf current technology to assist institutions and communities to enhance their connectivity, whereas what we are doing is research which will result in advanced technology. So we don't see any conflict there either.

The other part of the problem is one of scale. The amounts of money that are requested to be appropriated to NTIA are probably enough to do some good demonstration and prototyping work, but they are nowhere near enough to wire up the whole country.

REA, of course, has a small program as well that is of very great interest to us. So we meet with them and try to stay current with them.

Mr. BOUCHER. So you see NTIA's program as being compatible and supportive of the efforts that you have.

Dr. LINDBERG. Complementary, yes.

Mr. BOUCHER. Complementary.

I have two additional questions, and we have gone on quite a while. I will try to be brief with both of these.

We are getting ready to go to conference with the Senate shortly on the new applications program for the high performance computing program. The House version of that is H.R. 1757, which you very strongly supported, and we appreciate your help with that. Do you have any advice for us as we take this matter up with the Senate? Is there any particular set of changes that we should ask for in the provisions of H.R. 1757?

The Senate, you probably know, has a much thinner approach to this that just basically says carry out research, period, and we are somewhat more specific about the kind of research we would like to see conducted. We structured our approach following a number of hearings in which the interested public essentially told us the kind of research they would like to see done, and your agencies contributed to that process.

Are we basically on the right track, or have we learned something in the roughly one year since we passed that bill in the House that ought to argue for changes in the structure of the bill? Any advice that you have for us on that.

Dr. LINDBERG. I don't know if I am speaking for the group or not, but we like the provisions of your bill. The Senate version actually got less specific as it evolved. I am not certain why. I thought it was a very well written bill to start with. This committee added the fifth element of public access to Government information, which certainly was very well received by all.

I guess it is probably more important to get some bill through and get everybody behind it.

Mr. BOUCHER. Given your preference, you would take H.R. 1757 to the Senate counterpart. Is that fair to say?

Dr. LINDBERG. I don't know what the OSTP would say on that.

Mr. BOUCHER. Okay. I think I hear what you are saying.

Dr. Nelson.

Dr. NELSON. I can only speak for the Department of Energy on this. We worked very hard with other agencies and with OSTP on some language in S. 4 having to do with data communications, and the one piece of advice if I were asked—and I guess I have been asked—is to be a little careful in H.R. 1757 and in conference not to inadvertently create Federal procurement law.

We had some problem in the Department of Energy with provisions in an early version of S. 4 which, if interpreted one way—and you never know until it is law—might restrict our ability, in fact, to provide network services to our DOE laboratories, and we felt that that was making procurement law.

Mr. BOUCHER. All right.

Further comments?

Mr. Holcomb.

Mr. HOLCOMB. I would just comment that I think many of the provisions of H.R. 1757 have already influenced our program. You have heard today a number of the application areas. Access to Government information, in fact, is being pursued both within and without the Government but not necessarily fully within HPCC. Also, applications for remote sensing and K-12 education. These areas that are contained within H.R. 1757, have already influenced us. I think that by itself is extremely important, and I think we should recognize that.

Mr. BOUCHER. All right.

Any further comments?

The final question that I have is more technical in nature. Section 208 of the High Performance Computing Act requires that the Congress receive an annual report that is to be prepared by the Office of Science and Technology Policy that indicates the extent to which HPCC program awards have been received by entities outside of the United States, by foreign entities. It is supposed to be an annual report. To date we have not received any report. The program has been in effect for three years, and the administration has been silent with regard to this measure.

Dr. Lindberg, we have received some questions at this subcommittee from other Members of the Congress about why the HPCC program has not responded to this requirement. Do you have any advice for us as to why that has not happened?

Dr. LINDBERG. I have to offer you my apologies. We simply slipped up on that. We have now provided a copy of a report to the OSTP as of May 2, and so I am sure it will be acceptable to them and be forwarded to you. But I discussed it last night, and, with your permission, I will just give you the results; they are very brief.

Mr. BOUCHER. Okay.

Dr. LINDBERG. There are two elements to be reported, but to simplify it, one is support of research overseas, and so for the years 1992, 1993, and 1994 only ARPA—ARPA was the only participating agency that entered into any such activity using HPCC funds, and that activity is with the University College, London, for approximately $120,000 a year as part of the NREN Interagency International Program.

The second part to be reported was on essentially procurement of gear over $1 million that is manufactured overseas on overseas components, and for the years 1992, 1993, 1994, the answer is zero, and I apologize for not having done this. Our solution in the future is to just make this a part of the annual report. So it won't happen again, I assure you.

Mr. BOUCHER. That is fine, and thank you for the information. That is extremely helpful.

I would like to offer each of you the opportunity for the period of about the next three weeks to supplement the record of this hearing with any further statements you would care to make about the kinds of successes that have been achieved to date through the HPCC program, and I would encourage you to make those additional submittals if you are of a mind to do so, because they will assist us as we explain the benefits of the program in this public forum.

The other thing that I would ask is that if you have any recommendations for us beyond those mentioned here today for H.R. 1757 and the Senate counterpart as we go to conference with the Senate, that you share that with us as well.

And with the subcommittee's thanks, we will conclude this hearing. This has been highly informative. I would personally like to commend each of you for the excellent work you are doing. I think you have well highlighted the progress that is being made, and it is substantial progress indeed. As a strong supporter of the effort, you certainly have my congratulations and I think that of the American people as well. Thank you very much.

[Whereupon, at 11:23 a.m., the subcommittee was adjourned.]

APPENDIX

Supplement to May 10, 1994 Testimony before the Science Subcommittee of the House Science, Space, and Technology Committee by representatives of the HPCC Program

At the May 10 hearing, Rep. Boucher, Chairman of the Science Subcommittee, invited HPCC Program representatives to supplement their testimony before the Subcommittee with a report describing the major successes of the HPCC Program to date. The representatives thank Mr. Boucher for his invitation and submit the following report, which describes overall Program accomplishments and individual agency successes.

The document entitled, "High Performance Computing and Communications: Technology for the National Information Infrastructure" (commonly referred to as the Blue Book) is a Supplement to the President's Fiscal Year 1995 Budget that itemizes six major Program successes. These accomplishments, brought up to date, are as follows:

1. More than a dozen high performance computing research centers are in operation nationwide. More than 100 scalable high performance systems are in operation at these centers. These include large scale parallel systems, vector parallel systems, hybrid systems, workstations and workstation clusters, and networked heterogeneous systems. The largest of these systems now provides almost 150 gigaflops (billions of 1floating point operations per second) performance on benchmarks. The HPCC Program's 1996 goal of developing high performance systems capable of sustained teraflops (trillions of floating point operations per second) performance is well on the way to being met.

2. The Internet, seeded by HPCC funding, continues to experience phenomenal growth in size, speed, and number of users. This network of networks (with the HPCC-funded networks providing the initial core) enables researchers to access high performance computers and advanced scientific instruments easily, and has begun to allow independence of geographical location.

3. One illustration of the global reach of HPCC technology is that the Internet now extends across the country and throughout much of the world. The Internet links nearly 3 million computers, more than 18,000 networks in the U.S. and more than 13,000 outside the U.S., and 1,200 4-year colleges and universities, 100 community colleges, 1,000 high schools, and 300 academic libraries in the U.S.

4. More than half a dozen gigabit testbeds conduct research in high capacity networks. These network testbeds connect 24 sites, including many of the high performance computing research centers. Seven

Federal agencies, 18 telecommunications carriers, 12 universities, and two state high performance computer centers participate. The testbeds develop technologies to handle the NII's (National Information Infrastructure) increased demand for computer communications, along with greater accessibility, interoperability, and security. The Program goal of demonstrating gigabit (billions of bits) per second transmission speeds by 1996 is well on the way to being met.

5. Teams of researchers are using scalable systems to discover new knowledge and demonstrate new capabilities that were not possible with earlier technologies. These researchers are addressing the Grand Challenges, fundamental problems in science and engineering with broad economic and scientific impact whose solutions can be advanced by applying high performance computing techniques and resources. Many of these challenges are associated with HPCC agency missions. Agencies are increasingly working with U.S. industry to use their Grand Challenge applications software and develop new software that will improve commercial and industrial competitiveness.

   Users of the new scalable computing systems with their high performance on large problems and larger memories are able to address more complex, more realistic problems. We are beginning to understand our world better and to improve our lives:

   - Improved modeling of the Earth and its atmosphere has sharpened our ability to predict the movement and characteristics of storms and other forms of severe weather. With improved forecasts, there will be more lives saved and an overall positive economic impact due to reduced property loss and evacuation of smaller endangered areas along the coast and elsewhere.

   - New air and water quality models are enabling improved environmental decision making.

   - Improvements in the design and manufacture of goods are yielding better products. Both the production processes and products such as cars and airplanes are becoming more energy efficient.

   - We are learning more about how the human body functions and are improving our ability to diagnose and treat diseases.

6. The thousands of researchers who develop fundamental HPCC and NII technologies and applications form the vanguard that hardware and

software vendors rely upon to promote the use of high performance computers and communications throughout the U.S. economy. Hundreds of teachers and thousands of students access HPCC resources, and the Program conducts hundreds of training events for thousands of trainees. Dozens of small systems have been installed at colleges and universities. The goal of these efforts is to train a nation of knowledgeable users and thereby fully incorporate HPCC and NII technologies and applications into the U.S. economy.

In addition to these multi-agency successes, participating agencies have had notable successes both individually and jointly in developing and applying HPCC technologies. We asked each agency to provide some examples:

1. Advanced Research Projects Agency (ARPA)

Overview

During the past decade, ARPA investments in information technologies enabled fundamental advances that have been a major factor in the continued leadership of the U.S. in this key area. Previous ARPA investments in information technology tended to focus on relatively narrow technology sectors such as timesharing, computer graphics, computer networks, personal computing, and associated systems software and early applications. Each of these has had a fundamental effect on defense, commerce, and education. ARPA investments of the past decade have built upon and jointly leveraged all of these, and have transformed them into a new information technology that is also transforming society.

These results have recently become visible in the form of the Federal HPCC Program and its extension to include the technologies and pilot applications for the NII. Dramatic advances in these technologies have produced a new scalable information technology base that can be configured over a wide range of computing and networking performance, and distributed in the form of national- and global-scale systems.

New scalable technologies provide a combination of higher performance for the larger systems, with greater cost effectiveness for the smaller systems. In addition to the new technologies, a new generation of people who understand the technologies, as well as the new applications, has begun to emerge. The scalable nature of the technologies, coupled with the emergence of a pilot information infrastructure using Internet technologies, enables the accelerated diffusion of ideas, methods, technologies, and human resources.

Beyond these accomplishments, the Internet (used as a pilot information infrastructure) is transforming the way in which the information technology research community operates, the way it collaborates with industry, the way the scientific and engineering communities work with each other, and the way in which even broader sectors operate. To see this, one need only look at the explosive growth of the Internet, the number of users, computers, networks, and information servers, and the distribution of these systems by geography, by discipline, or by sector. These technologies and their application have considerably influenced Vice President Gore's National Performance Review, and are now forming the foundation for dramatically transforming the way the Federal Government operates into a new form of "electronic government."

Below we highlight successes in five major technology areas: (a) scalable computing systems, (b) microsystems, (c) scalable software, (d) networking, and (e) national-scale information enterprises.

1.a. Scalable Computing Systems

Over the past decade, ARPA invested in the development of a wide range of scalable computer architectures to explore the use of parallelism to increase performance. The technology innovations include processor architectures, cache and memory system architectures, interconnects, disk storage systems, and high performance communications network interfaces.

More specifically, ARPA pioneered the development of RISC-based computer systems, spanning workstations, shared memory multiprocessors, and large scale parallel architectures, including the associated operating systems and compiler technologies. The Sun SPARC and MIPS instruction set architectures can be viewed as direct commercializations of ARPA-funded research projects, and other architectures such as the IBM PowerPC are derived from these. The National Science Foundation (NSF) also played a key role in supporting the fundamental research that led to RISC concepts and systems.

ARPA provided support for demonstrating RAID (Redundant Arrays of Inexpensive Disks) systems with hundreds of small formfactor disk drives. This is now the architecture of choice for high performance, high capacity, high reliability I/O (input/output) systems, and has spawned commercial developments at over 20 companies, including Compaq, DEC, IBM, NCR, StorageTek, and Sun. NSF also provided research support for the early proof of the RAID concept.

A critical HPCC Program milestone was recently achieved when ARPA demonstrated 100 gigaops class systems as part of cooperative projects including Cray Research (T3D), IBM (SP1 and SP2), Intel (Paragon), Kendall Square Research (KSR 1), and Thinking Machines (CM-5). Many of these systems were first evaluated at NSF Supercomputer Centers. ARPA worked closely with vendors to accelerate developments in scalable architectures. For example, through a cooperative agreement with ARPA, the Cray T3D was delivered in 26 months on schedule and within budget. An early beta version of the machine was made available at NSF's Pittsburgh Supercomputer Center through joint NSF and ARPA sponsorship.

1.b. Microsystems

The design technology developed and validated within the microsystems program has created the advanced architectural building blocks for scalable parallel architectures. ARPA fueled the development of the computer-aided design software industry by supporting early and fundamental university-based research in integrated circuit layout, simulation, and synthesis tools. Many of the existing CAD software vendors can trace their roots to design software efforts that came from this program. Companies such as Cadence and Synopsis have built directly upon CAD software originally developed under ARPA sponsorship.

ARPA funded the technology developments that made it possible to demonstrate the first "virtual factory" integrated circuit process simulation shadowed by a real semiconductor fabrication line. State-of-the-art integrated circuit fabrication lines cost as much as $1 billion, making it imperative to validate the process prior to final implementation. The virtual prototyping techniques developed under ARPA sponsorship directly addressed this need by demonstrating approaches to optimize and tune the process before committing it to fabrication lines.

Scalable computer backplane communications technology provides the foundation for a majority of university and commercial scalable architectures. Work has been funded at Caltech, MIT, the University of Washington in Seattle, and others. Caltech's achievement in mesh routing has led to its adoption in various commercial systems, and the same components are now being exploited in high bandwidth, low latency, low cost local area networks.

ARPA funded the development of MOSIS (metal oxide semiconductor implementation service), the first network-based rapid implementation service for VLSI (very large scale integration) systems and an early prototype for NII-based design and commercial services. NSF worked with ARPA to make this service available to educators and students throughout the U.S., as well as to NSF's research community. This has led to a vibrant and well trained engineering community, able to capitalize on new technologies and to retain U.S. leadership. In addition, ARPA funded the development of FAST, the first-of-its-kind network-based broker for electronic and other catalog components. Time to order and acquire parts has been reduced by a factor of five to 10. FAST is being evaluated by the Defense Logistics Agency for possible operational use within DOD.

1.c. Scalable Software

ARPA's scalable software investments have led to many advances in the underlying technologies that support large, complex, and distributed applications. ARPA funded the development of microkernel operating system technology, which allows applications-level customization of operating system services for greater performance and flexibility. The microkernel software architecture forms the intellectual basis of the WindowsNT Operating System from Microsoft. The ARPA-developed Mach OS is the operating system of two major scalable computing vendors (DEC and Hewlett-Packard), and forms the foundation for continued innovations in OS.

ARPA developed the Computer Emergency Response Team (CERT) in response to the first large scale attack on Internet hosts. CERT has become the world leader in the fight against computer intruders. In addition to being the focus for intrusion detection and response, CERT has become a locus for creation and dissemination of protective software, tools for system security administration, and establishment of "best practice" for site computer security. ARPA and the National Institute of Standards and Technology (NIST) later collaborated to form the Forum of Incident Response and Security Teams (CERT-FIRST) to provide a government interface for CERT activities and extend the capabilities among Federal agencies.

The widespread deployment of the Internet has required significant advances in protection and trust mechanisms. ARPA has led in the development and deployment of security and privacy technology. ARPA developed privacy-enhanced mail for secure, trusted electronic mail interchange on open networks. ARPA, in cooperation with NIST and the National Security Agency (NSA), also developed the technology for Trusted Operating Systems. Through close coordination with developers of commercial versions of Mach, "Trusted Mach" or TMach, will inherit both performance enhancements and functional enhancements from its commercial variant. Another ARPA developed system, the ISIS distributed interprocess communication system, decreases the complexity of developing distributed applications by effectively eliminating most common distributed systems software failure modes. The strategies it employs are so effective that it has been commercialized by its developers and is being used for critical real time applications such as the American Stock Exchange trading system and the French Air Traffic Control System.

Acceptance and integration of Internet technologies in the commercial and military environments requires the ability to protect organizational systems. ARPA has supported the development of a suite of protective technologies for computer systems including public domain network firewall technology, controlled execution Unix, and trusted auditing tools. These tools enable systems administrators to increase the resistance of their systems to intrusion.

Libraries of high performance software executable over a range of heterogeneous machines are major enablers of scalable technologies. ARPA funded the development of SCALAPACK (the scalable version of the linear algebra package LAPACK), one of the most widely used libraries of mathematical software for high performance systems. This was a proof of concept demonstration that performance tuned mathematical software libraries enable applications to be written in machine independent fashion while achieving high efficiency on various high performance computer architectures. NSF supported the development of select portions of SCALAPACK.

ARPA also supported the development of new and extended programming languages for parallel machines, including FORTRAN-D, High Performance Fortran (HPF), Split-C, and data parallel and object parallel C++, as well as the associated advanced compilation technology for modern RISC and signal processing microprocessors. Compilers and compiler technology developed at Rice University under ARPA sponsorship have been transferred to IBM, Intel, Motorola, Silicon Graphics, Tera, and Texas Instruments. NSF also supports work in scientific compilers and languages through its Science and Technology Center at Caltech, Rice University, and Syracuse University.

1.d. Networking

Recognizing early the importance of computer communications, ARPA has invested in a variety of network developments over the past two decades. Initial funding was for ARPANET, the first packet-switched network and the predecessor of a broad industry of computer networks. ARPA funded the development of the Internet and its associated TCP/IP network protocol and, with the collective leadership of NSF and DOE, saw its evolution into the globe-spanning Internet. The Internet today includes more than 30,000 networks and 2,500,000 computers and enables communications amongst more than 20 million people around the world. More than half of today's Internet-connected organizations are commercial, and many of the Internet communication services are provided by commercial vendors. The development of TCP/IP was responsible for the creation of the network router industry that in 1994 is expected to have over $3 billion in sales. The TCP/IP protocol was recently adopted by Microsoft as the basis for all of its future networking products, which should result in complete penetration of the computer market. In addition, the ARPANET and its successor Internet have done more for cross discipline collaboration than any other program, activity, or incentive before or since.

ARPA developed the SURAN (Survivable Radio Network) family of networking protocols that permits reliable delivery of data in a hostile environment. This seminal work is a direct predecessor to the Cellular Digital Packet Data (CDPD) standards being created and deployed today. ARPA pioneered the development of packet-switched voice protocols in the 1970s and the packet video protocols in the 1980s, which provided the early proof of concept that packet technology would work for these media. This is the basis for the Asynchronous Transfer Mode (ATM) protocols used within the telephone system to integrate voice, video, and data. ARPA-sponsored contractors, such as Fore Systems, are among the world's leaders in supplying ATM switching technology for local area networks. In collaboration with the Department of Energy's (DOE) Lawrence Berkeley Laboratory, ARPA developed advanced video, audio, and shared whiteboard collaboration capabilities in conjunction with developing a multicast capability layered over the Internet. This multimedia conferencing capability is used regularly to disseminate information from conferences and other meetings as well as support informal meetings over the Internet, thereby further enhancing cross discipline cooperation.

In a strong collaboration with NSF, ARPA deployed five gigabit-class networking testbeds (described in section 2.f. below) and a sixth ARPA-only testbed. All six focus on network research as well as high performance applications. These testbeds have fostered interoperability among telecommunications service providers and have accelerated the availability of high bandwidth networking services to commercial customers. In addition, they demonstrated the need for gigabit-class or better network infrastructure for certain high performance applications such as three-dimensional radiation treatment planning and visualization.

1.e. National-Scale Information Enterprises

ARPA's leadership in network-based technologies and services yielded numerous early examples of NII capabilities within the ARPANET and later the Internet. ARPA pioneered the development, deployment, and operation of electronic commerce capabilities on networks for more than 10 years. These highly successful efforts include MOSIS, the silicon broker, and FAST, the semi-automated sourcing and buying system. Recently, FAST initiated a commercial pilot project with Citibank to interface the brokering system to a payment system, thus automating billing, credit authorization, and payment processing. FAST currently supports 135 customers and over 3,000 vendors.

In addition, ARPA has led the way in developing network-based tools for collaboration. These include the MBONE (multicast backbone) infrastructure supporting shared video over the Internet, as well as associated tools for shared design spaces.

ARPA has taken a pioneering position in digital libraries technologies by establishing early testbeds and working closely with the Library of Congress to prototype new publishing and recordation mechanisms suitable for the electronic world envisioned for the NII. In addition, ARPA's programs in scalable I/O, mass storage, and image processing systems all provide crucial technology that will underlie digital libraries.

As a specific recent accomplishment, ARPA supported MIT in developing the White House electronic document distribution system over the Internet. Over 2,000 documents have been distributed since January 20, 1993, to 4,500 direct subscribers and an estimated 120,000 end users. It is estimated that over 4 million page equivalents have been distributed in this fashion. This technology is a prototype of advanced subscription and dissemination capabilities needed for the NII.

2. National Science Foundation (NSF)

2.a. Supercomputer Centers

The four NSF Supercomputer Centers have become recognized leaders in the national HPCC effort. By providing reliable high performance computing services, they influence a broad range of scientific disciplines both within the academic community and across a wide range of industries, from computer vendors to producers of products for the general public. In this latter context, the Centers pioneered partnerships with the private sector to apply HPCC technologies and practices to specific design and manufacturing problems. Two illustrative examples are now described:

2.a.1. Cornell Theory Center - Merck Research Laboratory Collaboration:

The Cornell Theory Center (CTC), one of the four Supercomputer Centers, and Merck Research Laboratories (MRL) recently announced that the company has joined CTC's Corporate Research Institute (CRI) to explore scalable parallel computing technologies. A division of Merck & Co., Inc., the world's largest prescription drug company, MRL will be applying CTC's parallel computing resources to computer-aided drug discovery and structure determination. An interdisciplinary research center at Cornell University in Ithaca, NY, CRI was established to manage the exchange of knowledge and transfer of technology from CTC and its affiliated community to industry.

Dr. James B. Doherty, Executive Director of Biophysical Chemistry and Chemical and Biological Systems at Merck stated, "We are excited by the opportunity to interact with the scientists of the Cornell Theory Center as we explore the potential of scalable parallel computer systems. We believe such systems will play an important role in meeting the ever-increasing computational demands of computer-aided drug discovery. More importantly, we believe these technologies will be useful in our efforts to develop important new medicines."

Under the partnership agreement, Merck scientists have access to CTC's computational resources, most notably the IBM Scalable POWERparallel system (SP1), as well as to CTC consulting and training programs. CTC acquired the first SP1 in April 1993 with support from New York State. The SP2, a 512 processor system that will be capable of performing at 125 gigaflops (1/8 teraflops) speed, is scheduled for installation later this year. CTC has a strategic relationship with IBM to apply SP technologies to problems in disciplines such as fluid dynamics and rational drug design.

In addition to accessing CTC computing resources via network connections, MRL researchers and staff participate in CTC education and training programs. These programs range from an introduction to parallel computing to highly technical training in specific applications with hands-on experience.

Other CRI partners include Abbott Laboratories, BIOSym Technologies, Corning, and Xerox. CRI offers a rich variety of opportunities for its partners: a robust high performance computing environment, experimentation with a variety of highly parallel scalable systems, advanced algorithm research, and a strong record in interdisciplinary approaches to scientific and engineering problems critical to industry.

2.a.2. Mosaic

The second example of NSF Supercomputer Center activities is the freely distributed Mosaic software suite developed by the National Center for Supercomputer Applications (NCSA). Mosaic enables dynamic browsing, retrieval, and display of multimedia data (text, images, sounds, movies) from the World-Wide Web (WWW), an information retrieval system on the Internet. Mosaic uses hypertext to allow documents to contain words or images as direct links to additional information. Clicking on linked words or images retrieves and displays new files and information without having to know the location or network address of that information. Retrieved hypertext files in turn contain links to more information, creating a global web of information involving over 2,300 databases. In less than a year, Mosaic has become the fastest growing Internet-based application, with over 30,000 copies being delivered each month.

Recently, Spyglass Inc. and NCSA have entered into an agreement that will put Mosaic on the desktops of millions of people. The agreement gives Spyglass, which was formed in 1990 and has commercialized other NCSA technologies, full rights to enhance, commercialize, and broadly relicense Mosaic. Spyglass is making a multimillion dollar commitment to NCSA and will focus initially on developing a commercially enhanced version of Mosaic that other companies will incorporate in the products they distribute to their customers.

"We're committed to evolving Mosaic so it becomes a robust, commercial tool with complete documentation, technical support, and advanced features," explained Tim Krauskopf, cofounder of Spyglass. "We'll be collaborating with NCSA and other key partners to create new tools and establish standards that will help organizations build robust World-Wide Web information servers quickly and inexpensively."

Spyglass has redesigned Mosaic so it will be a more robust and full featured tool. Mosaic enhancements available from Spyglass include improved installation, better memory management, increased performance, new forms capabilities, online hypertext-based help, support for a proxy gateway, and user interface improvements such as support for multiple windows. Future versions will include enhanced security and authentication, which will enable credit card and other business transactions to take place on the Internet; filters that will enable documents from popular document readers to be read seamlessly by Mosaic; and integration with emerging editing and document management tools.

A number of businesses are already using Mosaic and WWW to publish magazines, deliver goods and services, provide technical support to customers, and conduct other forms of business electronically. For example, Mosaic and WWW are part of the recently announced $12 million CommerceNet project, a public- and private-sector-backed initiative exploring ways to conduct commerce over the Internet and other networks.

2.b. Grand Challenge Applications Groups

NSF, with ARPA support, funds 16 Grand Challenge Applications Groups pursuing solutions to scientific and engineering problems whose computational demands go beyond those provided by today's most powerful computing systems. Some highlights are summarized below.

2.b.1. Geophysical and Astrophysical Fluid Dynamics

One part of the NSF Grand Challenge Application Group effort in Geophysical and Astrophysical Fluid Dynamics at the University of Colorado and at the University of Minnesota is simulating turbulence in a computational cube of the finest resolution possible in order to understand the fundamental mechanisms driving turbulent flows. One goal is completing a series of calculations on a computational grid of one billion processors, that is, in a cube with a thousand computational zones in each of the three dimensions. In order to be practical, such a calculation requires on the order of 28 gigabytes of addressable memory and many hours of dedicated system time executing in the range of several gigaflops.

Professor Paul Woodward, a member of the team from the University of Minnesota, presented the requirements to Silicon Graphics (SGI) as a possible demonstration of that company's high performance computing capability. The result was a collaboration in which SGI agreed to put together a network of high end workstations configured specifically to support the execution of Woodward's piecewise parabolic method hydrodynamics software on a grid of one billion computational zones. For the Grand Challenge team, questions of compelling scientific importance could be explored; for its part, SGI could demonstrate that a cluster of their workstations could attack one of today's largest computational problems.

Over a three week period, SGI configured 16 of its Challenge XL systems with 20 processors each, and in collaboration with the Minnesota team, addressed specific software issues that had to be resolved in order to conduct a successful simulation. The periodic boundary conditions used in the simulation necessitated interconnecting the systems in a 2x2x4 three-dimensional toroidal array using FDDI (fiber distributed data interface) network technology. Thus, the experiment represents one of the few times that the scientific application requirements drove computing system design; normally, a manufacturer provides a system and users must adapt their algorithms to that system's design characteristics.

The numerical simulation was carried out over the space of a week using the full system at 30 percent of peak efficiency including communication time overhead. This resulted in a sustained computation rate of 4.9 gigaflops. It will take much longer to analyze the output and describe the findings. To facilitate this process the results were written on 350 4-gigabyte tapes organized to support the analysis as well as restarting the calculations from specific points if necessary.

This billion-zone turbulence computation demonstrated that workstation computing involving hundreds of processors must be taken seriously by the high performance computing community. It was made possible by the arrival of a new generation of workstations, represented by the SGI Challenge XL Servers used here, which have very fast interconnections of tens of powerful new microprocessors with multiprocessing software tools to match. Simultaneously, "off the shelf" high speed interconnection networks now can produce highly effective arrangements of tens of these workstations with relatively little effort or cost. Additionally, high speed disk and tape peripherals can now be attached to this new generation of workstations. The effort also demonstrated the power of some of the extensions to the Fortran programming language that the Minnesota team has developed and used in their simulations. SGI is investigating the use of these extensions in their product line as well as their possible impact on standards efforts such as the High Performance Fortran Forum, which focuses on developing a portable Fortran that supports parallelism.

2.b.2. Understanding Human Joint Mechanics through Advanced Computational Models

At Rensselaer Polytechnic Institute, Professor Rober Spilker leads a group of engineers, mathematicians, and computer scientists in an effort to develop realistic computational models of human skeletal joints using actual anatomic geometries and multiphasic properties of joint tissues. Early results focused on the role of cartilage in joint movement. In particular, accurate models for cartilage deformation and for the observed time-dependent and velocity-dependent friction coefficient of cartilage have been developed. These models will be used to better understand the interaction between cartilage and the interstitial fluid in joint motion.

2.b.3. High Performance Computing for Learning

Researchers at MIT under the direction of Professor Robert Berwick seek to understand how a person or a machine can learn visual, motor, language, and other skills, thereby laying the foundation for a future generation of high performance computers that include powerful brain-like functions. The approach is to combine computational models with biologically driven learning models in order to design VLSI chips that enhance machine learning. For example, a novel memory-intensive VLSI chip has been demonstrated as a PC "coprocessor" that delivers high performance computer performance on the face recognition problem.

2.b.4. Earthquake Ground Motion Modeling in Large Basins

Professor Jacob Bielak leads a team of researchers centered at Carnegie Mellon University who are developing the ability to predict the ground motion in large basins during strong earthquakes, with particular attention to better understanding the Greater Los Angeles Basin. To this end the team is building tools to help the application specialists solve mechanics problems on parallel computers. One such tool is Archimedes, a special purpose compiler that supports the mapping onto parallel systems of unstructured grid computations required in basin modeling. Such discipline-specific tools demonstrate that non-specialists can make effective use of parallel systems without becoming intimately involved with the idiosyncrasies of a particular architecture.

2.c. Science and Technology Centers

NSF supports four Science and Technology Centers whose missions are directly related to the HPCC Program: The Center for Cognitive Science at the University of Pennsylvania, the Center for Computer Graphics and Scientific Visualization at the University of Utah, the Center for Discrete Mathematics and Theoretical Computer Science at Rutgers University, and the Center for Research in Parallel Computation (CRPC) at Rice University. These centers support a broad program from basic research to technology transfer to educational programs aimed at exposing K-12 students to the excitement of science.

CRPC conducts basic research in parallel languages, compilers, and programming. It is a leader in the High Performance Fortran language specification effort that is supported by most parallel systems vendors. It has formed strong ties with industrial and scientific applications in order to provide feedback on the relevance of the research developments. For example, the Geosciences Parallel Computation Project involves participants from a number of universities and the petroleum industry focused on using parallel computation to simulate enhanced oil recovery methods. Study areas include flow in porous media, seismic analysis, and optimal well placement in the context of developing advanced tools for parallel scientific programming.

2.d. Education and Training

NSF supports a variety of efforts aimed at K-12 through graduate education and postgraduate and industrial training. The Supercomputer Centers and the Science and Technology Centers play a major role in this activity. The SuperQuest Program provides a specific example.

SuperQuest is a nationwide summer program for high school students and their teachers. Each SuperQuest entry team has teacher-coaches and three or four students who work on computational science projects that are judged in a competition among all submitted projects.

The student finalists and advisers attend a three week educational and training seminar that involves "hands-on" experience with a variety of high performance vector and parallel computers, including executing applications software related to their particular science projects. These seminars are held at designated SuperQuest sites, each of which hosts four winning SuperQuest teams.

Students and teachers attend classes in subjects ranging from advanced mathematics to English, including training in effectively presenting results to an audience. They also benefit from associating with academicians, industrial researchers, mentors, and consultants at the sites. The teams return home to continue their computational research, where they use advanced workstations awarded to their schools by equipment vendors through the SuperQuest program. Each team also receives one year of Internet connectivity so it may continue to access the computers and communicate with the researchers at the SuperQuest sites.

In 1993 the SuperQuest sites were CTC, NCSA, Reed College, and the University of Alabama at Huntsville; in 1994 the University of North Carolina, Chapel Hill, will be added. Similar programs have been initiated in Arizona, California, Minnesota, and New Mexico.

2.e. Institutional Infrastructure — Minority Institutions Program

NSF provides grants to computer and information science and engineering departments in minority institutions (usually African American or Hispanic) to aid in research, education, and outreach activities. The program supports equipment acquisition and Internet access in order to lower barriers to full research participation by minority institutions. The education component requires that the university increase minority participation in computer science and engineering by upgrading faculty qualifications and undergraduate research participation. The outreach component has two parts: (1) active recruitment of entering freshmen through visits to high schools and K-12 participation in science projects, and (2) cooperative agreements between minority and majority institutions in order to provide opportunities for graduate work for students from the minority institutions.

For example, the University of North Carolina A&T, an historically black university, supports both research and education and is addressing the shortage of African Americans in computer science and engineering through its Communications Methods Research Laboratory, Signal Processing Laboratory, and VLSI Design Laboratory. The school is linked to Duke University, Michigan State University, Stanford University, and the University of Michigan as preferential graduate schools for NC A&T graduates. Under this program, NC A&T has provided 30 minority students to the Ph.D. pipeline and strengthened its research programs.

2.f. Gigabit Testbeds

NSF and ARPA jointly fund five gigabit network testbeds led by the Corporation for National Research Initiatives (CNRI) for research on very high speed communications networks and Grand Challenge applications. Testbed collaborators come from universities, national laboratories, high performance computer centers, and major industrial organizations that greatly leverage the Federal investment. The testbeds include three wide area networks, AURORA, BLANCA, and CASA, and two metropolitan area networks, NECTAR and VISTANet. Researchers are exploring issues important for developing and using more widely deployed gigabit networks, including network architectures; interfaces between the network and high performance computers, graphics engines, and workstations; high speed and real time protocols; distributed computing; shared memory; stream-lined, network-friendly operating systems; traffic and network control; and application software environments. The testbeds are used to explore Grand Challenge applications that cannot be investigated without the combined use of resources linked by gigabit networks, including high performance computers, graphics engines, high performance workstations, and sensors as varied as MRI scanners and astronomy observatories. These applications include climate modeling that combines ocean and atmospheric models, medical imaging, multimedia digital libraries, multiple remote visualization and control of simulations, radiation treatment planning, and radio astronomy imaging.

3. Department of Energy (DOE)

DOE and its predecessor agencies and commissions have a long history in high performance computing, both in defense programs and in unclassified energy research programs. The National Energy Research Supercomputer Center (NERSC) was established in 1975 to provide for fusion energy's high performance computing needs, and was made available to all energy researchers in 1984. DOE also established the predecessor of the Energy Sciences Network (ESnet) to provide access by NERSC users, the majority of whom were located off site. ESnet is now an integral part of the Internet.

DOE was one of the founders of the HPCC Program. It participates in all components of the Program but emphasizes the computational applications. The HPCC Program enabled DOE to establish two new HPCC Research Centers, dramatically improve ESnet, undertake eight Grand Challenge Projects, move its important applications to parallel processing through advanced parallel algorithm development, and initiate educational programs in computational science. Many of the most significant DOE accomplishments are described in HPCC reports to Congress. Several more recent accomplishments are described below.

3.a. Sandia/Intel Shatter Computing Speed Record

A team from Sandia National Laboratories and Intel achieved a world record 143.4 double-precision gigaflops on the Massively Parallel LINPACK benchmark. The previous record of 124.5 gigaflops was set in August 1993 by Fujitsu's Numerical Wind Tunnel, a one-of-a-kind system owned by the Japanese National Aerospace Laboratory. A key technology enabling this performance was SUNMOS, the Sandia UNM (University of New Mexico) Operating System. SUNMOS uses only two percent of each processor's memory and operates at peak communication bandwidth.

"Sandia's achievement certainly validates the soundness of Intel's scalable architecture and the leading role that Sandia National Laboratories plays in pushing the envelope of high performance computing," said Ed Masi, president of Intel's Supercomputer Systems Division and vice president of Intel Corporation. "The LINPACK record represents a win not just for Sandia and Intel, but for the U.S. Advanced Research Projects Agency, the Department of Energy and the entire U.S. HPCC community."

3.b. Technical Conferencing Capabilities

With the increasing size and the geographical diversity of scientific collaborations, the role of high speed networking in supporting collaborative work must be expanded. Under the HPCC Program, computer scientists at the Lawrence Berkeley Laboratory work with scientists at Xerox Palo Alto Research Center to develop tools for network-based collaboration and conferencing.

Such collaboration has three basic needs: an audio tool, a video tool, and a shared workspace or "whiteboard." In addition, network protocols needed a number of modifications in order to support video conferencing. In order to provide good voice transmission, a time stamp was added to each packet to compensate for the variable time delay in network communication. A multicast protocol was developed in order to minimize the bandwidth requirements of video transmission. This protocol ensures that only one copy of any video conference is transmitted on a single network link. When the video signal reaches a point where it must follow two or more separate links, multiple copies of the conference are generated by the multicast routers.

Used together, these tools have become an invaluable part of ESnet and the Internet. Beginning with the networking and computer science communities, use of conferencing tools has now spread to many DOE science programs and beyond. On April 15, 1994, the first Multicast Symposium in the medical community was held at the Long Beach Regional Medical Education Center at the Long Beach VA Medical Center in California. This Pancreatic Islet Micro-circulation Symposium brought many experts in the field together to establish and summarize the current state of knowledge of islet microcirculation. Most of the conference was broadcast directly from the Center but one presenter, Susan Bonner Weir of Harvard, was unable to attend in person and gave her presentation remotely from the MIT Laboratory of Computer Science. The symposium's success demonstrated the important role that video conferencing tools can play in the medical community.

3.c. Simulation of Grinding Process in Virtual Reality

Researchers in Argonne's Mathematics and Computer Science Division are using the CAVE (Cave Automatic Virtual Environment) to gain new insights into the grinding process, an essential process in manufacturing. This is collaborative work with Argonne's Reactor Engineering Division and with industrial scientists at Surface Finishes Co.

Grinding a component involves several controllable parameters such as materials properties, rotational speed, surface roughness, and force/interface between the block and the grinding wheel. Virtual reality enables the researcher to better understand the effects of these parameters on the grinding process. Using CAVE tools in conjunction with Argonne's large scale parallel IBM SP1 computer, researchers observe temperatures and stresses within the block and wheel. Sounds of the surface motions are played through the CAVE audio system, ablated material is ejected as small particles and displayed as sparks, and the researchers have tactile feel of the dials and knobs that control the process.

Argonne and its collaborators envision several benefits from the CAVE simulations: (1) information gleaned from the virtual reality environment can aid in the development of new manufacturing processes, (2) the CAVE/SP1 integration will enable researchers to explore alternative designs in real time, and (3) virtual reality will facilitate training in performing manufacturing operations.

3.d. HiPPI/SONET Gateway: Paving the Information Superhighway

As a result of the HPCC Program, several high performance local area computer networks that use gigabit network technology based on the High Performance Parallel Interface (HiPPI) have been built across the country. This is the highest performance local area networking infrastructure commercially available today. The Los Alamos HiPPI interconnects the CM-5, Cray T3D, High Performance Data System (HPDS) (described below), SGI visualization high performance computers, plus various workstations and workstation clusters. While these local area networks can reach up to a few kilometers, interconnecting testbeds that are farther apart was a critical need.

With HPCC support, Los Alamos developed the HiPPI/SONET Gateway to solve this distance limitation. This Gateway uses SONET (Synchronous Optical NETwork), which is the commercial infrastructure now being implemented by local and long distance telephone carriers. The Gateway uses SONET 155 megabit channels (OC-3C) to obtain the full HiPPI bandwidth by "striping" seven of these channels to achieve the full 0.8 gigabit/second bandwidth along with forward error correction.

The Gateway is accomplishing two important milestones: (1) the link installed in August 1993 between Los Angeles and San Diego holds the world speed record of 660 megabits per second between two widely separated high performance computers, and (2) the installation in the second quarter of 1994 of the link between Los Alamos and San Diego. These links, especially the latter, will be the fastest in operation among the gigabit testbeds. The Los Alamos link will also have the highest bandwidth - distance product of any testbed now in operation.

3.e. Global Climate Modeling and the CHAMMP Program

Global climate simulations offer the best scientifically grounded framework for studying our changing planet. The HPCC Program together with the DOE CHAMMP (Computer Hardware, Advanced Mathematics, and Model Physics) program has been instrumental in defining a climate simulations development path that will use the power of large scale parallel processors. Oak Ridge National Laboratory (ORNL), in collaboration with Argonne and the National Center for Atmospheric Research (NCAR), has provided the first full implementation of a state-of-the-art atmospheric climate model on a large scale parallel computer. To explore the feasibility of using such computers for climate research, the Parallel Community Climate Model PCCM2 has been optimized and configured for parallel I/O. The high I/O requirements of climate models is currently driving the large scale parallel computer vendors to advance computing technology in this key area. One-year climate simulations have been performed on the ORNL Intel Paragon XP/S 35 to validate climate statistics in preparation for an even more accurate 10 year simulation using high spatial resolution. This simulation will explore the importance of fine scale processes in climate prediction. This simulation has not been performed on conventional high performance vector computers because of the high memory requirements and the dedicated computational resources required to complete the study.

3.f. The Los Alamos High-Performance Data System (HPDS)

Under the auspices of the HPCC Program, the implementation of the Los Alamos High-Performance Data System (HPDS) has resulted in significant technological breakthroughs in data access and data storage system science and has laid the groundwork for future data storage systems.

The HPDS was developed to meet the enormous data storage and data access requirements of DOE Grand Challenge problems, and is the first fourth-generation data storage system to be put into production. It can handle data generated by problems running on the most advanced, large scale parallel, large memory computers and high speed cooperative processing networks in the world. HPDS is currently being used extensively to handle massive amounts of data from global climate model simulations. In addition, knowledge and experience gained by implementing and using HPDS is being used in designing and implementing the High-Performance Storage System (HPSS), one of the largest collaborative efforts between national laboratories and industrial partners. Requests for HPDS have opened the door for future collaborations and technology transfer. Without HPCC support, HPDS development and implementation would not have happened, and data storage technology advancement would be substantially delayed.

3.g. Manufacturing for the Semiconductor Industry

Sandia's large scale parallel Direct Simulation Monte Carlo (DSMC) software is being used to simulate plasma etch and deposition reactor designs for the microelectronics industry. The software models fluid flow in the high vacuum regime where continuum equations break down. The software is being used extensively to model process physics under a Sandia/SEMATECH CRADA. Lam Research has already saved money by using the DSMC software to evaluate the parameter space for a new design; previous empirical tests had failed to meet the design goals. Companies including Applied Materials, IBM, Intel, Lam Research, and Motorola are working with Sandia to set up their own problems and have them executed on the Intel Paragon and nCUBE high performance computers at Sandia.

3.h. Domain Mapping Software Advances Parallel Computing

Sandia researchers have developed Chaco, a software package that provides reliable, automated methods for mapping problems onto large scale parallel computers. Efficient parallel execution requires that a problem be decomposed into equally sized subproblems that are assigned to processors in a way that minimizes data exchange between processors. Sandia researchers have recently made important progress in solving this problem. Combining graph theory, numerical analysis, and advanced computer science techniques, Chaco is a sophisticated software package implementing a variety of new and advanced mapping algorithms. It is designed to cover a broad cost/quality range, allowing users to choose methods that give reasonably good mappings very quickly or that provide nearly optimal mappings in moderate time. In addition, the new techniques are quite general and can be applied without expertise to many scientific computations.

Since its release in October 1993, Chaco has been licensed for use at more than 35 external sites, gaining rapid acceptance at universities, corporate research facilities, and other national laboratories.

4. NASA

Overview

NASA's HPCC program is an integral part of the Federal HPCC Program. It is structured to contribute to broad Federal efforts while addressing agency-specific computational problems that are beyond projected near-term computing capabilities. NASA's HPCC research and development work fully supports the Federal Program's efforts to dramatically improve U.S. capabilities in high performance computing and communications. The NASA program will enable the agency to demonstrate the testbed technology for a thousand-fold increase in computing performance and to apply that technology to NASA Grand Challenges.

The goals of NASA's HPCC program are to accelerate the development, application, and transfer of high performance computing technologies to meet the engineering and science needs of the U.S. aeronautics and Earth and space science communities, and to accelerate the implementation of the NII. These goals are supported by four specific objectives: (1) to develop algorithm and architecture testbeds that are fully able to use high performance computing concepts and increase overall performance, (2) to develop high performance computing systems scalable to sustained teraflops performance, (3) to demonstrate HPCC technologies on U.S. aeronautics and Earth and space science research problems, and (4) to develop services, tools, and interfaces essential to the NII.

NASA has Grand Challenges in Computational Aerosciences (CAS) and Earth and Space Sciences (ESS). These Grand Challenge applications were chosen for their potential and direct impact on NASA, their national importance, and their technical challenge.

4.1 Computational Aerosciences (CAS) Project

The Grand Challenge in Computational Aerosciences is to accelerate the development and availability of high performance computing technology of use to the U.S. aerospace community, to facilitate the adoption and use of this technology by that industry, and to hasten the emergence of a viable commercial market in which hardware and software vendors can exploit these advantages.

To support these goals, NASA has pioneered the implementation of design and simulation software on parallel systems. This marks the first implementation of aerospace simulation software on parallel systems, showing the way toward making them useful to industry. Additionally, NASA developed the most widely accepted performance evaluation and tuning software for applications on parallel systems; developed the only reliable set of benchmarks with which to judge the many parallel architectures; and provides the most complete set of testbed systems to the aeroscience community. Among its most important accomplishments, NASA has:

- Provided an entire suite of first generation parallel systems for evaluation by the aerospace community

- Led the implementation of aerospace design software on the new parallel systems at the NAS (Numerical Aerodynamic Simulation) facility (Intel Paragon with 208 processors and Thinking Machines CM-5 with 128 processors installed)

- Awarded a Cooperative Research Agreement (CRA) for the HPCC second generation parallel testbed and related research to a consortium headed by IBM

NASA recently enabled remote high speed network access to its testbed systems and successfully demonstrated aerodynamic and multidisciplinary simulation of HSCT (High Speed Civil Transport) configurations using large scale parallel systems.

4.2 Earth and Space Sciences (ESS) Project

The Grand Challenge in Earth and Space Sciences is to demonstrate the potential afforded by teraflops systems to further our understanding of and ability to predict the dynamic interaction of physical, chemical, and biological processes affecting the solar-terrestrial environment and the universe. The computing techniques developed as part of NASA's HPCC program will further the development of a suite of multidisciplinary models leading ultimately to scalable global simulation coupling many models, or for treating highly energetic multiple scale problems in astrophysics.

The Earth and space science community is interested in the ESS research achievements that relate key enabling computational science areas. Top space and Earth scientists are Principal Investigators (PIs) and collaborators on eight ESS Grand Challenge Teams and 21 Phase-1/Phase-2 Guest Computational Investigations selected by NASA Headquarters through the ESS NASA Research Announcement (NRA) process; this has established computational metrics for these teams. Using large scale parallel computing systems to conduct research has become routine for these investigators, but how best to transfer their experience to the broader scientific community has yet to be determined.

A new core ESS performance evaluation activity will produce ESS parallel benchmarks to expose underlying factors that determine performance. NASA has joined with NSF in this effort by forming the Joint NSF-NASA Initiative in Evaluation (JNNIE).

Working with Goddard Space Flight Center (GSFC), the ESS project has begun deploying ATM-based LANs within GSFC and connecting to DOD's (Department of Defense) local 2.4 gigabit/second ring around Washington, DC.

The ESS project has developed and demonstrated advanced parallel computational techniques in collaboration with ESS PI Teams. These are interdisciplinary critical mass teams that are developing complete Grand Challenge applications scalable to teraflops. Some of the critical scientific areas are:

- Scalable Hierarchical Particle Algorithms for Cosmology and Accretion Astrophysics. Dr. Wojciech Zurek at the Department of Energy's (DOE) Los Alamos National Laboratory (LANL) is the PI. The scientific Grand Challenge is to provide physical insight into astrophysical phenomena of cosmological significance through numerical modeling. The computational challenge is to improve the numerical statistics of the simulations and better represent the physical processes involved in controlling the evolution of the universe.

- Earth System Model: Atmosphere/Ocean Dynamics and Tracers Chemistry. Dr. Roberto Mechoso of the University of California at Los Angeles is the PI. The scientific Grand Challenge is to develop a global coupled model of the atmosphere and the oceans, including chemical tracers and biological processes, to be used to model seasonal cycle and interannual variability. The computational challenges are to make it possible to test the impact of changing parameter values in these models rapidly and to visualize time-accurate model output in real time.

- Large Scale Structure and Galaxy Formation. The PI is Dr. George Lake at the University of Washington. The scientific Grand Challenge is to allow a physically meaningful comparison between empirical data and hydro-dynamic models of large scale structure and galaxy formation.

The ESS project has also developed a teaching curriculum for parallel computational techniques for the annual NASA Summer School in High Performance Computational Physics at GSFC.

Further, the ESS project established the Center of Excellence in Space Data and Information Sciences (CESDIS) University Research Program in Parallel Computing and linked each research project with a major NASA activity. Under this project, ESS worked closely with Convex, Cray Research, Intel Parallel Systems Division, and MasPar, in developing their scalable parallel systems, and installed major testbeds from these manufacturers at Jet Propulsion Laboratory (JPL) and GSFC to make their systems (Convex SPP-1, Cray T3D, Intel Paragon, and MasPar MP-2 Cluster) available for evaluation by ESS investigators.

4.3 K-12 Project

The NASA K-12 Project supports the IITA's (Information Infrastructure Technology and Applications, a component of the HPCC Program) efforts to establish standards and provide working models for commercial communications infrastructure. NASA has developed a plan by integrating a combination of 33 schools, NASA HPCC K-12 projects, and Regional Training Centers, located in 14 states and the District of Columbia. The project's goal is to significantly contribute to NASA's role in meeting national goals for science and mathematics K-12 education in areas appropriate for high performance computing and communications technology. The objectives are to involve all NASA centers participating in the HPCC program, involve other centers as appropriate, obtain a broad national presence, impact early stages of education (middle school and earlier), inspire students with NASA mission applications, enable students to become electronic information explorers, and facilitate communication among separated communities to empower synergistic collaborations.

One example of project implementation is the current effort in the State of Maryland that provides connectivity to every school district within the State. This was accomplished through the Maryland/Goddard Science Teacher Ambassador Program. That Program's overall objective is to introduce secondary school Earth or environmental science teachers to NASA Earth and Space Science data, information, Internet tools and applications, for subsequent dissemination to students and other educators.

5. National Institutes of Health (NIH)

5.a. Visible Human Project — National Library of Medicine (NLM)

Images are an important part of biomedical knowledge, and high performance computing has enabled new methods for visualizing three-dimensional anatomic structure. With HPCC resources, NLM has undertaken the first steps toward building a collection of three-dimensional digital images of human structure at sub-millimeter level resolution. Named the Visible Human Project (VHP), this image collection is derived from Computed Tomography, Magnetic Resonance Imaging, and digitized cryosections of cadavers. As a reference standard for three-dimensional human anatomy, the VHP data will enable creation of "living" textbooks of anatomic structure that allow students and health care practitioners to view internal body structures from any viewpoint, using interactively generated three-dimensional imaging. Extending these methods to individualized treatment planning will provide an unprecedented ability to plan accurate and minimally invasive surgery.

5.b. Telemicroscopy — National Center for Research Resources (NCRR)

The increasing availability of high speed computer networks makes feasible a system in which a scientist working at a graphics workstation in a laboratory can perform a research investigation by deploying tasks to specialized scientific instruments and high performance computers located at remote sites and interconnected to the workstation over the network. The Microscopist's Workstation project explores such a "distributed laboratory" in which a researcher will be able to remotely acquire and process images of a biological specimen using a special electron microscope in combination with high performance computers. The project is designed to extend the availability of an intermediate high voltage electron microscope, a unique scientific laboratory instrument located at the San

84-115 O - 94 - 5

Diego Microscopy and Imaging Resource (SDMIR), through the use of advanced networking technology.

The microscope can image considerably thicker biological specimens with greater contrast than more commonly available lower voltage instruments. Images of thick specimens provide important information about the specimen's three-dimensional structure. The microscope is being used to study several cellular phenomena, including the disruption of nerve cell components in Alzheimer's disease.

The microscope is controlled by a local workstation connected via high speed networking to the remote researcher's workstation. Network voice and data links allow the investigator to communicate with the microscope operator and to collect images in a variety of forms in near real time. These images can be routed via the network to a high performance computer for automatic processing into a digital volume representing the interior of the specimen.

The Microscopist's Workstation project is a collaborative effort among the SDMIR at the University of California San Diego (UCSD) School of Medicine, the San Diego Supercomputer Center, the Scripps Research Institute, and the UCSD Department of Computer Science and Engineering. Costs are being shared by NCRR, NIH, NSF, the State of and University of California, and industrial partners (Datacube, Network Systems, and Sun).

5.c. Predicting Molecular Structure From Genetic Code — National Cancer Institute (NCI)

Large scale parallel computation and high speed network communications have become important parts of the research effort at NCI's Laboratory of Mathematical Biology. An optimized and parallelized version of the Zuker RNA folding software, which predicts molecular structure by analyzing the linear sequence of RNA (the messenger molecule produced by genes), has been adapted to a MasPar MP-2 large scale parallel computer. The program is able to analyze RNA sequences that are over 9,000 nucleotides in length; this is substantially larger than the maximum length that can be analyzed by the Cray YMP computer. The HIV virus is an example of such a sequence. NCI researchers have demonstrated that the MP-2 can fold such sequences 4.5 times faster than a single processor of the YMP, thus giving a considerable price/performance advantage. Efficiently folding such large sequences will enable a detailed comparison of structural motifs that may be common in viral families such as HIV, and support both vaccine development and targeted drug design.

6. National Institute of Standards and Technology (NIST)

6.a. Guide to Available Mathematical Software (GAMS)

A wealth of reusable software is freely available for use by scientists and engineers for solving recurring types of mathematical and statistical problems. As high performance computers become more widely available, it is increasingly important that this high quality software be easily accessed and used. Unfortunately, the average user has difficulty locating the most appropriate software within the volume of specialized software distributed over computer networks.

The NIST Guide to Available Mathematical Software (GAMS) project develops techniques and tools to help scientists and engineers locate and use computer software to solve mathematical and statistical problems. The products of this work include a problem classification scheme, a database of information about available software, and an information retrieval system for user access. The information system functions as a virtual software repository; rather than maintaining a repository itself, it provides a convenient cross-index of repositories managed by others. Moreover, GAMS uses Internet resources to connect to these repositories and obtain source code, documentation, and tests for its users. For example, some 1,600 software modules from the well-known netlib repository have been cataloged. GAMS retrieves these items from the netlib server at ORNL. The complete GAMS database contains information on more than 9,200 software modules from some 60 software packages at four physical repositories.

Anonymous telnet access to the GAMS system is available to Internet users. Both a generic and a sophisticated X11 networked graphical user interface are supported. Some 1,700 connections from about 150 Internet hosts were logged during its first eight months of (largely unadvertised) operation. Recently, a new network access mechanism was developed based on NCSA Mosaic. GAMS availability via Mosaic has had a strong impact on user access — more than 17,000 GAMS Mosaic access requests were recorded during April 1994.

The GAMS project supports the goals of the ASTA component of the HPCC Program in facilitating the widespread reuse of algorithms and software developed for high performance computers. It depends upon Internet advances in order to effectively deliver services to its users and on Mosaic software developed by NCSA under NSF HPCC funding. GAMS is one of the software exchange mechanisms being studied and coordinated by NASA.

6.b. The MultiKron Performance Measurement Chip

The basic measure of computer performance is the length of time required to complete a task. But then comes the question, 'Why did it take that long?" The answer is particularly difficult to obtain in parallel computers, where problems are broken into many parts and solved simultaneously by many processors. This contrasts with traditional computers where problem parts are solved sequentially by one processor.

Scalable parallel computing systems constitute a growing market encompassing both vast machines at the high end of computing and desk-side personal workstations. Yet the technology is still immature. Large scale parallel programming is not easy and system features are not firmly established. Different programming techniques are required by each parallel architecture.

The NIST parallel processing group investigates performance measurement methods that promote scalable systems. Accurate measurement permits understanding the source of performance bottlenecks and therefore leads to techniques to scale system designs upwards. This approach addresses concerns in achieving the economic promise of affordable large scale parallel systems.

In cooperation with ARPA, NIST has designed the MultiKron chip technology for measuring the performance of advanced processors and very high speed networks. MultiKron technology achieves low-perturbation data capture for assessing the performance of scalable parallel computing systems. This technology was transferred to Intel, as well as to other companies and researchers. Intel applied the MultiKron measurement principles in its scalable parallel Paragon system, whose designers and users can obtain accurate performance information and use that information to achieve maximum throughput.

NIST also designed, developed, and distributed a designer and experimenter's kit for using the industry standard VME bus to integrate MultiKron instrumentation into systems, and used MultiKron instrumented systems to assess the performance of TCP/IP communications protocols over a HiPPI-based network.

7. National Security Agency (NSA)

7.a. SPLASH 2 System Development

Researchers at the Supercomputing Research Center (SRC) of the Institute for Defense Analyses, a federally funded research center, and NSA have developed SPLASH 2, a general purpose hardware accelerator board that can deliver high performance computer processing performance for a variety of applications at greatly reduced cost. In the most basic terms, SPLASH 2 provides a collection of raw logic gates that can be programmed through software. The primary processing elements of SPLASH 2 are Xilinx 4010 field programmable gate arrays (FPGAs). Each FPGA is connected to a 256K x 16-bit RAM. Communication between the processing elements is performed via a linear data path and a crossbar interconnection network. The linear data path enables SPLASH 2 to perform systolic algorithms, while the crossbar connection makes it possible to implement algorithms that require irregular communication. The combination of these two communication patterns facilitates the data transfers required for a diverse set of applications.

A SPLASH 2 system consists of a Sun SPARCstation host, an interface board, and between one and 16 SPLASH 2 array boards. SPLASH 2 is programmed in the VHSIC (Very High Speed Integrated Circuit) Hardware Description Language (VHDL), which is an IEEE and DOD standard programming language. This language allows users to configure the system hardware to perform specific processing functions. Applications implemented in this manner are able to run at the speed afforded by hardware while maintaining the flexibility provided by software. Thus SPLASH 2 is an excellent computing platform for researchers who do not have the resources to design and build hardware for every new high speed processing task they wish to perform.

Several applications have been implemented on the 10 existing SPLASH 2 prototype systems. Researchers at Virginia Tech are using SPLASH 2 to process image data at real time video rates. Image processing algorithms that have been successfully implemented on SPLASH 2 include edge detection, Fourier transforms, image convolution, and histogram generation. Work is scheduled to begin at Michigan State University that will use SPLASH 2 for fingerprint identification at rates that are orders of magnitude faster than those achieved using current methods. SPLASH 2 algorithm research is not confined to universities. The National Cancer Institute is using SPLASH 2 to test newly identified DNA strands against an existing database so that these new strands can be evaluated and categorized. Although these applications begin to demonstrate the variety of tasks that can be efficiently executed on SPLASH 2, other applications in signal processing and prototype development environments may also benefit from its processing capabilities. Future work will concentrate on refining the SPLASH 2 hardware and making the technology available to researchers for examining these unexplored application areas.

7.b. ATM/SONET Project

Over the past two years, the NSA Information Security Research organization has successfully addressed the problem of incorporating encryption in ATM and SONET networks. This has consisted of defining the methods of operation and building and demonstrating proof-of-concept encryptors to operate in test versions of these networks. To define the systems, NSA has been working with the vendor community, standards groups, and users at the national and international levels to create command structures, protocols, and formats that allow efficient use and management of encryption in these environments. NSA is also developing the necessary vendor independent interfaces between host communications equipment and the encryptors.

This program addresses the secure use of future common carrier communications systems. ATM is the technology with the most promise for these applications. Unlike present carrier services where capacity is allocated and billed regardless of actual information transfer, in ATM the capacity of a physical link is allocated only as needed based upon demand for transferring information. This is the concept of a "virtual circuit" — the user appears to have an allocated circuit, but the network only transfers information for the user when information is present. This efficient sharing of network capacity potentially represents dramatic cost savings to telecommunications users. Because of the seemingly random mixing of information flows along a communications path, each packet of information potentially belongs to a different virtual circuit. Any encryption concept, therefore, has to switch between user states at the packet rate of the network. This "Key-Agile" encryption allows multiple ATM virtual circuits/channels to operate on a per demand basis, in conformance with the principles of ATM virtual circuit communications, while maintaining confidentiality between the various sets of users without degrading network capacity. The proof-of-concept models have focused on the ATM networks, but the same basic concepts are baselined into KG-189, for securing SONET links, and into FASTLANE, a production encryptor for securing ATM virtual circuits. Verifying the validity of the operational concepts in a realistic environment is an important part of this program. NSA is a node on ATDnet (Advanced Technology Demonstration network), DOD's Global Grid Washington-area ATM/SONET testbed, which is presently completing installation and checkout. Once stable, ATDnet will be used to test and demonstrate NSA's prototype encryption equipment, thereby paving the way for fully operational and approved encryption equipment.

7.c. TERASYS Development and Cray-PIM Supercomputer

The Terabit Operation Per Second (TERASYS) project goal is to demonstrate large scale parallel computing in the memory of a general purpose computer. The system is based on a specially designed Processor-In-Memory (PIM) chip containing 64 processors. Five TERASYS workstations were produced during the past year. Each contains 512 PIM chips and computes at approximately the same speed as an eight-processor Cray YMP. A software environment has been written to make the hardware ready for user-written programs.

The next step is incorporating PIM chips into the Cray-3 computer. If successful, this will result in a prototype of a commercial PIM high performance computer. Cray Computer would then have the expertise to build production machines. Using existing memory density technology, the prototype is expected to achieve a speed equivalent to 200 to 800 Cray YMP processors at a fraction of the cost of current high performance computers. This machine would be suited for parallel applications that require a small amount of memory per process. Examples of such applications are laminar flow over a wing, weather modeling, image processing, and other large scale simulations or bit-processing applications. Because it is based on memory technology, PIM technology is readily adaptable to future memory density advances.

8. National Oceanic and Atmospheric Administration (NOAA)

NOAA is a mission agency responsible for describing, predicting, and understanding changes in the Earth's environment. To perform and to advance this mission, NOAA develops and applies leading-edge computation and communications technologies to perform Grand Challenge research in climate and weather modeling, predict the weather, understand and predict ocean processes, and manage the Nation's fisheries.

To advance its capabilities in high performance computing, NOAA has acquired a Cray C90 high performance system for its National Meteorological Center (NMC, Suitland, MD) for operational weather forecasting. It has acquired a 208-processor Intel Paragon for its Forecast Systems Laboratory (FSL, Boulder, CO) to be used as a state-of-the-art testbed for developing advanced algorithms in climate, ocean, and weather modeling. This acquisition was accomplished in conjunction with NOAA's partners in the ARPA-sponsored National Consortium for High Performance Computing and as part of the Boulder Front Range Consortium. NOAA is conducting a major technological upgrade to its high performance computer facility at the Geophysical Fluid Dynamics Laboratory (GFDL, Princeton, NJ) to support research in numerical algorithms for environmental modeling.

8.a. High Performance Computing

NOAA has long been a major developer and user of advanced computational technologies in Grand Challenge applications. Over the past several years, NOAA has had major successes in advancing our understanding of the environment through the use of advanced high performance computing technologies. Among NOAA's recent accomplishments are:

- The Parallel Ocean Program (POP), a variant of the GFDL Modular Ocean Program (MOM), has been successfully transported by scientists at DOE/Los Alamos National Laboratory (LANL) and GFDL, to the LANL 1,024-processor Thinking Machines CM-5. Scientists are now using this model to perform very high resolution experiments to produce multi-year eddy-resolving simulations of the world oceans.

- As part of the NOAA/DOE collaboration, GFDL's SKYHI global atmospheric grid-point model (used to study climate, middle atmosphere dynamics, and atmospheric ozone) has been thoroughly redesigned and transported to the LANL CM-5. GFDL and LANL scientists are now preparing to conduct global atmospheric experiments with half degree horizontal resolution (producing 50 kilometer grid spacing over the United States) to investigate the effect of small scale features on stratosphere and mesosphere dynamics. These very high resolution experiments require computing resources that are beyond the capabilities of the current GFDL computer.

- NOAA scientists, collaborating with DOD/Naval Research Laboratories (NRL) scientists, have successfully implemented a version of the NMC atmospheric spectral model to the NRL Thinking Machines CM-200.

- A regional atmospheric model has been restructured by NMC and FSL scientists for execution on large scale parallel systems. NOAA's FSL scientists have developed a parallel version of the Mesoscale Analysis and Prediction System (MAPS) as a functional prototype for both the Federal Aviation Administration and the National Weather Service. MAPS is the first of several strategic weather models to be parallelized for the Aviation Weather Program. These models will provide high resolution forecasts on both national and regional scales to support operational and aviation meteorology.

- NOAA's Pacific Marine Environmental Laboratory (PMEL) has entered into a cooperative agreement with the Arctic Region Supercomputing Center (ARSC) of the University of Alaska, which will provide PMEL and other NOAA scientists with access to ARSC's high performance computing resources, including a 128-processor Cray Research T3D. This collaboration will allow PMEL scientists to incorporate new parallel computing capabilities into their research models for ocean circulations of the Gulf of Alaska and the Bering Sea.

8.b. Environmental Information Dissemination through Networks

In performing its mission, NOAA collects, archives, and disseminates environmental data and information. It operates three National Data Centers and more than 30 additional major environmental information archive and distribution facilities. To fulfill this mission, NOAA has taken a pioneering role in advancing the use of the Internet and the envisioned NII to open its vast archives of environmental data and information for easy public access. NOAA is applying new information discovery tools that have emerged from the HPCC Program, such as Mosaic, to its environmental information systems to provide ready user access across the diversity of NOAA information. HPCC technologies play a strategic role in allowing NOAA to provide an integrated view across geographically distributed information systems so that a virtual "NOAA environmental information system" can be developed.

NOAA's recent accomplishments in providing timely, universally accessible information services include the following:

- NOAA now has 29 information servers on the Internet providing a wide variety of real time and historical environmental data and information to the world.

- As of the latest count, NOAA has implemented 15 Mosaic servers that access major NOAA databases, thereby providing simple, highly graphical, multimedia access to a broad range of environmental information.

- Using the new information technologies, NOAA has developed the ability to provide scientists, policy makers, educators, and students with immediate access to information that, until recently, would have taken weeks or months to find and retrieve from archives and distribute to the requestor. NOAA's information servers connected to the Internet provide convenient, instantaneous access to centuries of global weather observations, the current status of solar X-rays and particles as they are detected by NOAA's Geostationary Operational Environmental Satellites (GOES), current and recent past winds and sea temperatures across the tropical Pacific used to monitor El Nino events, and a wide range of other atmospheric, geophysical, and oceanographic information.

- A prototype Satellite Active Archive system connected to the Internet provides access to vast quantities of near real time NOAA satellite imagery as well as to historical satellite imagery.

- The NOAA Data Directory Services allows Internet users to browse information about the agency's environmental data holdings including information about applicability for a variety of uses and how to obtain the data.

9. Environmental Protection Agency (EPA)

9.a. Environmental Modeling

High performance computing has enabled the linkage of air and water models to address the problem of nutrient overloading in the Chesapeake Bay that can cause dramatic decreases in the productivity of the shellfish industry. Atmospheric deposition is an important pathway for nitrogen and toxic chemicals, possibly accounting for 25 to 35 percent of the nitrogen going into the Bay. Through the use of new technology and optimized computer software, EPA scientists have been able to use nitrogen oxide deposition estimates from an air quality model as input to water quality models to determine how reductions in emissions of nitrogen oxides translate into changes in nutrient load in the Bay High performance computing has created possibilities for greater cost efficiency in solving the problem because, beyond a certain point, air controls may be relatively less expensive than water controls. HPCC has enabled a more complete cross-media examination of options, allowing decision makers to take advantage of added benefits to regional ozone control by reducing air emissions of nitrogen.

High performance computing has facilitated the use of advanced air quality models in a retrospective study of the costs and benefits of the 1970 Clean Air Act. The use of advanced models for this type of assessment was typically infeasible due to computational limitations. High performance computing allowed more pollutants to be modeled, bringing in regional acid deposition, sulfate, visibility on an annual basis and regional ozone on a seasonal basis, while at the same time enabling uncertainty analysis and model evaluation to ensure scientific credibility of the results. This had never been done before using a single, self-consistent modeling system.

As part of efforts to develop an environmental decision support system, researchers at MCNC have demonstrated the use of dependency graphs for automated execution of multiple dependent programs including meteorology and emissions processing for the Urban Airshed Model. The processing graphs are created using the Heterogeneous Network Computing Environment (HeNCE), a public domain integrated graphical environment for creating and running parallel programs over heterogeneous computers. MCNC also completed a portable suite of modules, data, and networks for an Applications Visualization System (AVS)-based analysis component for the environmental decision support system. Linkages between AVS and ARC, from Environmental Systems Research Institute, were constructed to facilitate visualization and three-dimensional animation of data in geographical information systems.

A multidisciplinary team at Carnegie Mellon University has developed a Geographic Environmental Modeling System (GEMS) designed to facilitate in-depth study of the environmental effects of air pollution. GEMS provides a way to explore the relative effectiveness of alternative emissions control strategies. It is transportable to a number of computer systems, accesses data remotely via servers, and uses a graphical user interface for ease of use.

Researchers at the University of Wisconsin have demonstrated the applicability of three-dimensional animation techniques for air quality models. New volume rendering and interactive data selection techniques were developed for VIS-5D (VISualization of 5-Dimensional data sets) to accommodate the extensive data requirements of air quality models.

9.b. Computational Techniques

Researchers at Caltech have implemented six advection algorithms on both sequential (IBM and Sun workstations) and parallel (a network of IBM RISC workstations and a 256-processor Intel Touchstone Delta) platforms and performed comparative testing. Advection is a major component of air quality models. Results indicate that the accurate space derivative method coupled with the Forester filter produces the most accurate results, although at the expense of more computation time. However, a speed-up of about 88 is achieved with the parallel implementation on the Intel.

A gas-phase chemistry quasi-steady state approximation solver has been transported to a 32-processor Kendall Square Research KSR 1 by researchers at North Carolina State University. Numerous comparative studies indicate that this algorithm is more suited to vector processors. Experiments with other algorithms indicate the importance of optimizing local cache performance. The energy portion of a quantum mechanical model of large molecules has also been transported to the KSR 1 by researchers at Duke University.

As part of a joint research effort between EPA and DOE, researchers at Argonne National Laboratory have implemented a parallel version of the Mesoscale Meteorology model (MM5) using a portable message passing library on an IBM SP1 and an Intel Delta. Almost 2 gigaflops was achieved using 120 processors on the SP1 and about 3 gigaflops was achieved using 480 processors on the Delta. These are encouraging results given the importance of meteorological models for air quality simulation.

Four new cooperative agreements for research on parallel algorithms for ground water flow and transport, water quality, aerosol chemistry, and prediction of reactivity and structure of organic and biochemical compounds in solution are being awarded in FY 1994.

EPA's Scientific Visualization Center (SVC) completed tests of distributed visualization techniques involving all available computing resources at the Research Triangle Park and Bay City Visualization Centers. Capabilities include access to large remote data sets, distributed parallelization of computationally intensive tasks, time sequenced visualization, animation, and recording. These techniques are now being incorporated into the overall visualization infrastructure to support agency-wide distributed visualization.

EPA has established a desktop video conferencing capability between scientists at Research Triangle Park and the EPA's National Environmental Supercomputing Center (NESC) in Bay City, Michigan using MBONE (multicast backbone) digitized video techniques over the Internet. This link is used routinely for planning and status meetings, training sessions, and user support for software optimization and visualization. These capabilities are being extended to additional EPA remote sites.

9.c. Network Connectivity

EPA established a FDDI research network in Research Triangle Park, NC with a high speed (T3 or 45 megabits per second) Internet connection to NESC. T1 (1.5 megabits per second) connectivity has been provided to the Environmental Research Center in Georgia and to the Chesapeake Bay Office in Maryland and agreements are being established to extend network connectivity to the Ozone Transport Commission in New York and the Tennessee Valley Authority. This connectivity provides the bandwidth needed to satisfy heterogeneous computing, data management, and collaborative visualization requirements for EPA's HPCC research and technology transfer pilot programs.

9.d. Education/Training

A workshop was held for Federal, state, and industrial personnel participating in technology transfer pilot programs to begin an assessment of training needs and user requirements for use of air quality decision support systems. A prototype regulatory modeling and analysis support system, UAMGUIDES, was demonstrated for use in urban ozone non-attainment studies and is being provided to all participants for their continued use and evaluation. Multimedia electronic "tutor and help" approaches, televisualization, as well as collaborative approaches to provide expert advice to remote locations are being evaluated.

EPA's HPCC program helps support EarthVision, a computational science educational program for high school students and teachers. Saturday tutorials at NESC provide basic knowledge and skills required to prepare project proposals. Competitively selected teams participate in a three week summer education program. During the academic year, the teams conduct environmental research using scientific workstations and EPA's high performance computer.

EPA researchers have been given hands-on visualization training, and a student summer research program provides undergraduate opportunities for summer employment involving HPCC activities. Numerous undergraduate and graduate students and a postdoctoral student are supported directly through cooperative research agreements with universities.

HPCC

HIGH PERFORMANCE COMPUTING AND COMMUNICATIONS

FY 1995 Implementation Plan

April 1994

National Coordination Office for HPCC

EXECUTIVE OFFICE OF THE PRESIDENT
Office of Science and Technology Policy

HIGH PERFORMANCE
COMPUTING AND
COMMUNICATIONS

National Coordination Office for High Performance Computing and Communications

National Library of Medicine • 8600 Rockville Pike • Bethesda, Maryland • 20894
Phone: (301)402-4100 • FAX (301)402-4080 • E-mail: hpcc@nlm.nih.gov

April 8, 1994

John H. Gibbons, Ph.D.
Director
Office of Science and Technology Policy
Executive Office of the President
Old Executive Office Building
Washington, DC 20500

Dear Jack:

I am proud to convey to you the Implementation Plan covering FY 1995 for the High Performance Computing and Communications initiative.

This document covers in detail the plans the agencies have formulated to implement the overall objectives of the initiative. These objectives were embodied in the FY 1995 Annual Report entitled "High Performance Computing and Communications: Technology for the National Information Infrastructure" (the Blue Book). The Blue Book accompanied the President's budget, with your endorsement and that of the Committee on Information and Communications of the National Science and Technology Council.

Several elements of the Implementation Plan are different this year from its predecessors and are, to my mind, improvements. The detail with which the budgetary commitments have been described is substantially greater than before. The organization of the material is according to a format of Accomplishments and Objectives that was worked out with considerable OMB guidance and help. This document covering up through FY 1995 budget will be made available to the public.

On behalf of all the HPCC agencies, the fine professionals who serve in these agencies, and the many helpers and advisors from industry, academia, and the public, please accept my thanks for your leadership of this effort.

Respectfully yours,

Donald A. B. Lindberg, M.D.
Director

Enclosure

HPCC FY 1995 Implementation Plan

Table of Contents

Page

List of Tables

High Performance Computing and Communications Program

FY 1995 Implementation Plan

1. Executive Summary

This document contains the implementation plan for the High Performance Computing and Communications Program (HPCC) for FY 1995. The plan is based on the President's request of $1.154 B for HPCC. The HPCC Program is carried out by 10 federal agencies (ARPA, NSF, DOE, NASA, NIH, NSA, NIST, NOAA, EPA, and ED). This plan covers both HPCC-base activities that were in place prior to the beginning of the official HPCC Program in FY 1992 and activities begun as a part of the Program's new funding. To date, the HPCC Program has accomplished extraordinary results relative to the goals set out in 1992. This plan focuses on the efforts needed to continue to meet those goals. The HPCC process for reporting its plans continues to evolve; this report is the first time material is being disseminated at this level of detail. HPCC is planning to adopt this approach, with periodic updates for the future.

This plan also covers a recently added component activity (IITA - Information Infrastructure Technology and Applications) to provide the basic research and technology development that will help contribute to making the Administration's National Information Infrastructure initiative a reality. This plan identifies the new activities for IITA and adjusts the previously defined four HPCC components to support these additional goals and objectives.

Computing and communications are ongoing, strategic enabling technologies. Timely progress in these technologies

- Expands scientific and technical knowledge and capabilities,
- Enhances U.S. industrial competitiveness, and
- Improves the quality of life for the general public.

The current federal HPCC Program provides additional stimulation and coordination to accelerate progress in critical areas of the field.

The success of the HPCC Program is measured both in terms of tangible numerical metrics as well as less tangible influences on the direction of industrial practices and the general public. Examples of the former include sustained computer performance, network communication bandwidth and deployment, and data storage capacity, and indeed these are used by the Program to gauge progress in the developing technologies. But they do not tell the whole story. A less obvious benefit of the Program is the way it has accelerated progress by encouraging the early use of advanced information technology. This has helped industry to maintain its competitive edge and has enhanced the public's well-being. HPCC will continue its success as long as it provides significant enabling technology improvements that are used by industry and the public.

To date, the HPCC Program has shown considerable progress, both in terms of numerical metrics and "in-use" measures. In the High-Performance Computing Systems (HPCS) component, several companies (supported in part by HPCC) have demonstrated 100 gigaops class modules and systems. In the broader picture, the computing industry and users have recognized that high performance systems of the future can and will be scalable; this recognition represents a significant change in the commonly accepted wisdom in less than three years.

The National Research and Education Network (NREN) component continues rapid expansion in the number of site connections and use of the Interagency Internet. Over 100 new connections to the Interagency Internet were supported in FY 1993, including medical and public health institutions, educational institutions, and agricultural extension offices. (Connections to the entire Internet system are growing at over 10 percent per month.) Adding connections is well ahead of schedule, at least for services at T1 and lower speeds. Six gigabit testbed sites are operational, and deployment of high-speed network facilities has begun (up to 2.4 gigabits/sec.). Testbed teams consist of government agencies, universities, computer companies, and telephone companies that participate both as service providers and experimenters. Possibly the best measure of success of NREN is the current recognition that an "Information Superhighway" is a reachable goal for general public use.

In Advanced Software Technology and Applications (ASTA), 34 Grand Challenge scientific teams are developing applications on high performance computer systems. HPCC is also funding over a dozen high performance computer centers, serving over 12,000 users. New software tools are getting into users' hands, including message passing libraries, a parallel linear algebra library, 3D visualization tools, and performance monitoring and instrumentation toolkits. For example, these ASTA efforts enable much more detailed study of phenomena that affect everyone. A new hurricane model in 1993 accurately predicted 48 hours in advance and continuously thereafter, that Hurricane Emily would turn sharply back out to sea off Cape Hatteras without making landfall.

The Basic Research and Human Resources (BRHR) component continues to produce critical research results along a broad spectrum of topics (e.g., modeling molecular dynamics, RNA structure prediction, coherent laser sources, correctness proofs for VLSI processor design, parallel numerical algorithms, and virtual reality). BRHR also expands educational opportunities and training for all academic levels (e.g., SuperQuest, EarthVision, the National High School Honors Program, Adventures in Supercomputing, outreach to historically black colleges and universities, graduate assistantships, and post-doctoral research grants). A number of major universities have acknowledged the long-term importance of high performance computing by instituting cross-disciplinary Computational Sciences curricula. These studies train people to use computing effectively to enhance research efforts in other scientific disciplines. Based on the results to date, the HPCC Program is well on its way toward meeting many of its original goals.

Nevertheless, the HPCC Program has identified areas that need to accelerate progress to keep the program on schedule. This Implementation Plan includes adjustments to address these points. In HPCS, performance analysis techniques are not yet adequate to assess how various applications will scale up from small prototype machines. ARPA and NSF will continue to deploy prototype systems and stress efforts to find better performance prediction techniques within universities. For computationally intensive problems, agencies will begin evaluation and development work on larger machines that can be more readily expanded into full production systems. In NREN, efforts to upgrade the speeds of various links of the Interagency Internet have gone more slowly than originally planned, and some planned new awards for gigabit research have been delayed. Challenges to a NASA/DOE procurement have delayed one network effort almost two years. A recent NSF plan for network improvements is also being challenged. In addition, less than requested HPCC funds to DOE has increased delays in achieving network speed milestones. This Implementation Plan reaffirms the importance of the networking component and revises milestone timetables. In ASTA, efforts are expanding to establish stronger ties with independent software vendors. The objective is to reduce the risks and costs associated with moving applications that are critical to industry onto scalable, high performance computing systems.

The HPCC Program's original focus of enhancing the Nation's computing and communications capabilities now extends to address a broader set of technologies and applications that have an immediate and direct impact on critical information capabilities affecting everyone. The IITA component will produce enabling technologies important to developing the National Information Infrastructure. It will demonstrate feasible solutions to information intensive aspects of problems of national importance, such as health care and agile manufacturing, using the potential of the nation's high performance communications and information processing capabilities. The IITA component activities will :

- Create the underlying scalable computing technologies for advanced communication services and applications support services that can adapt to the available resources;
- Create an intelligent service layer that will significantly broaden the base of computer information providers, developers, and consumers while reducing existing barriers to accessing, developing, and using advanced computer services and applications; and
- Develop a more effective software development paradigm and technology base.

The new IITA component builds on and expands the research goals of the existing HPCC components. HPCS will integrate the IITA developments in real-time transaction processing, storage services, highly-reliable software, and mechanisms for enhanced privacy and security into a scalable systems context. To respond to the IITA needs for universal network services, new capabilities in NREN will provide far greater numbers of connections, higher volume traffic, greater redundancy needs, more diversity among the communication bitways (e.g., satellite, wireless, and cable), and more asymmetry to the communications media (e.g., using cable to broadcast services to customers, but using phone lines for return responses). ASTA efforts will continue to focus on developing specialized software tools, environments, and libraries for high performance computing applications while leveraging wherever possible the new IITA developments in the areas of software for secure, reliable communications, user-friendly interfaces, interoperable software building blocks, and tools for the access and use of large diverse data sets. BRHR will expand specific areas of basic research in support of all of these previously mentioned topics.

The extension of the HPCC Program to incorporate the IITA objectives is well underway. ARPA, NSF, NASA, NIH, and NIST are building substantial efforts in IITA. Examples include:

- A prototype distributed file system has been deployed enabling access to files across the Internet as an extension to the user's local file system (ARPA);
- The Mosaic software (for discovery, retrieval and display of multimedia documents) now has over one million public users, about 1,000 information servers available throughout the Internet, and is available in the public domain for a variety of common laptops, workstations, and high performance graphics workstations (NSF);
- A joint research initiative for digital library research and prototypes has been defined and requests for proposals have been issued (ARPA, NSF, NASA);
- A solicitation for cooperative agreement proposals has been issued to provide public access to national databases of remote sensing imagery over the Internet (NASA);
- Advanced telemedicine networking projects have begun in California, Illinois, Indiana, Iowa, Kansas, Oregon, Pennsylvania, and West Virginia (NIH); and

- Advanced manufacturing and electronic commerce integration and demonstration facilities are under development, with industry participation being solicited (NIST).

However some HPCC agencies have not received funding for IITA, thus delaying their contributions in certain areas. For example, only minimal efforts in energy demand and supply management (DOE) and dissemination of environmental monitoring information (NOAA, EPA, DOE) can begin in FY 1995. Also, although the technology is expected to be available, electronic dissemination of high resolution satellite imagery, environmental warnings and forecasts, spatially referenced environmental data, and in situ and remote sensing data will be delayed. In the areas of secure and authenticated communications, DOE and NSA have key capabilities to contribute to: secure information resources discovery and retrieval, intrusion detection, multi-realm cross authentication, smart card technologies, secure electronic commerce, secure distributed engineering and design environments, and biometric identification evaluation.

To achieve its objectives, the HPCC Program will continue and expand its interactions with the Administration's National Science and Technology Council. Its Committee on Information and Communication (CIC) has oversight responsibility for the HPCC Program. The HPCCIT Subcommittee (responsible for coordinating agencies' activities in the HPCC Program) will work closely with CIC to enhance oversight activities related to HPCC (including efforts to name an HPCC Advisory Committee).

2. Introduction

The High Performance Computing and Communications (HPCC) Program was formally established following passage of the High Performance Computing Act of 1991 (Public Law 102-194), signed on December 9, 1991. Ten federal agencies in collaboration with scientists and managers from U. S. industry, universities, and laboratories have developed the HPCC Program to meet the challenges of advancing computing and associated communications technologies and practices. *High Performance Computing & Communications: Toward a National Information Infrastructure* [1] describes the vision of the HPCC Program, defines the major organizational components and elements, and highlights key accomplishments that have been achieved to date. This plan provides a detailed description of the agencies' HPCC implementation plans for FY 1994 and FY 1995[2].

Since its inception in 1992, the HPCC Program has undergone changes that make it difficult to track actual funding against the originally planned funding profile. New agencies have been added, major new responsibilities have been added, and some agencies' activities not originally defined within HPCC have now been moved under this funding category. In FY 1991 before the formal start of the program, the estimated HPCC-related activities among the original eight agencies[3] totaled $489.4 M. The original funding profile for the HPCC Program began with this base level and the Program goals justified a level of funding that was approved to grow to roughly $1,200 M over five years. The current funding profile for HPCC shows:

- $638 M, FY 1992 Actual
- $728 M, FY 1993 Actual
- $936 M, FY 1994 Congressional Action[4] (This funding includes new responsibilities for IITA and adds two agencies[5] to HPCC. These agencies had FY 1993 HPCC-related activities totaling $43.6 M)
- $1.154 B, FY 1995 Presidential request.

This Implementation Plan contains three additional sections. Section 3 provides an overview of the HPCC Program definition and organization. Section 4 contains a breakdown of the five major components of the HPCC Program, with an emphasis on the overall directions and milestones planned for each one. Section 5 provides a detailed look at HPCC Program activities within each agency. Although the Department of Education is an official HPCC agency, its current funding and reporting of crosscut activities goes through the Committee on Education and Health Resources, not the HPCC Program. For this reason the Implementation Plan covers nine HPCC agencies.

---

1 A 1994 report by the Committee on Physical, Mathematical, and Engineering Sciences Federal Coordinating Council for Science, Engineering, and Technology, Office of Science and Technology Policy.

2 This document assumes the reader is generally familiar with the HPCC Program vision and definition. For the most part, it only briefly summarizes the material found in the above referenced 1994 report.

3 ARPA, NSF, DOE, NASA, NIH, NIST, NOAA and EPA

4 FY 94 estimated funding is approximately $ 157 M less than the President's FY 1994 request.

5 NSA and ED

3. Program Overview

The best source for getting a current overview of the HPCC Program is the 1994 document, *High Performance Computing & Communications: Toward a National Information Infrastructure*[6]. For purposes of explaining details of the HPCC Implementation Plan, portions of that material are included here. This section provides a broad overview of the definition, organization, and management used to implement the HPCC Program. The HPCC goals and objectives provide the focus for the global management, planning and organization of activities. Through a process of collaborative coordination, the participating agencies seek to leverage each other's activities wherever possible and to minimize redundancy in programs. However, funding for the HPCC Program flows directly to each agency participating in the Program. Each agency has its own mechanisms and approach to select and evaluate projects funded under this Program. A wide variety of methods are used to distribute the results of HPCC activities and to evaluate the overall progress of the HPCC Program.

3.1. Goals and Objectives

The High Performance Computing and Communications Program goals are to:

- *Extend U. S. technological leadership in high performance computing and computer communications;*
- *Provide wide dissemination and application of the technologies to speed the pace of innovation and to improve the national economic competitiveness, national security, education, health care, and the global environment; and*
- *Provide key enabling technologies for the National Information Infrastructure (NII) and demonstrate selected NII applications.*

These goals will be realized by accelerating and focusing the HPCC activities within each of the federal agencies charged with carrying out this program and coordinating those activities among the participating agencies. Many critical applications in industry, national security, and the environment require far greater computing capability than is currently available. The HPCC Program will accelerate the development of scalable computing systems that will have the capability and capacity to address more of these critical applications. It will also accelerate development of the critical supporting technologies, such as file storage systems, computing environments and network communications required to support the effective use of these computers. The Administration's vision for a National Information Infrastructure places unprecedented demands for network connectivity, capacity, database availability, information management, access security, and ease of use. The HPCC Program will work with industry to help to create important elements of the technology base needed for a universally accessible NII and will use this technology to develop and demonstrate prototype "National Challenge" applications. All of these activities depend heavily on the development of more cost-effective approaches to writing and maintaining software and the development of a larger pool of highly-trained professionals to use these computers and networks. The HPCC Program will develop and deploy software, tools, and improved algorithms to meet these programmatic needs. It will also establish sufficient human resources of educators, scientists, and other trained professionals in computational science and engineering to permit effective use and application of the HPCC technologies.

---

6 This document is also available on-line (World Wide Web and Gopher access) with other relevant information.

The HPCC Program is organized into five components with the following key aspects:

High Performance Computing Systems (HPCS):

- Accelerated development of scalable computing systems, with associated software, including networks of heterogeneous systems ranging from affordable workstations to large scale high performance systems
- Technology advances in component packaging, mass storage, and communications that support the design and use of scalable computing systems

National Research and Education Network (NREN):

- Broadened network connectivity of the research and education communities to high performance computing and research resources
- Accelerated research, development, and deployment of networking technologies

Advanced Software Technology and Algorithms (ASTA):

- Prototype solutions to Grand Challenge problems
- Improved algorithms, software technologies, and software tools for more efficient use of scalable computing systems
- Deployment of advanced high performance computing systems for R&D

Information Infrastructure Technology and Applications (IITA):

- Prototype solutions to National Challenge problems using HPCC enabling technologies
- Accelerated development and deployment of NII enabling technologies

Basic Research and Human Resources (BRHR):

- Support for research, training, and education in computer science, computer engineering, and computational science; and infrastructure enhancement through the addition of HPCC resources

3.2. HPCC Management, Planning and Organization

The High Performance Computing and Communications Program is implemented as a partnership among federal agencies and other organizations. The HPCC Program reports to the Director of the Office of Science and Technology Policy (OSTP) At the direction of the Director, Program oversight and budgetary review are provided by the Committee on Information and Communication (CIC) of the National Science and Technology Council. The National Coordination Office (NCO) for High Performance Computing and Communications provides a central focus for the Program. The Office coordinates the activities of the agencies, other organizations participating in the Program, and organizations providing services to it. The Office also provides an interface to Congress, industry, academia, and the public. Donald A. B. Lindberg, the NCO Director (and concurrently Director of the National Library of Medicine), reports to John H. Gibbons, the Director of the Office of Science and Technology Policy.

The High Performance Computing, Communications, and Information Technology (HPCCIT) Subcommittee handles planning for the HPCC Program. The Subcommittee meets monthly to coordinate agency HPCC Programs through information exchanges, the common development of interagency programs, and the review of individual agency plans and budgets. HPCCIT charters a variety of Working Groups to coordinate activities in specific areas, Task Groups to carry out specific limited duration studies, and Task Forces to make various planning recommendations. In most cases, individual agencies are assigned responsibility to lead these groups. Currently active groups include:

The Scientific and Engineering Computing Working Group, led by NASA, has the mission to identify, promote, transfer, and coordinate software developed under HPCC.

The NREN Group, led by NSF, coordinates network integration activities and works closely with the Federal Networking Council (FNC), which consists of representatives from federal agencies that use federally funded networks. The FNC is responsible for coordinating the efforts of government HPCC participants and other NREN governmental constituents, in addition to providing a liaison to nonfederal communities interested in the federal program. A Federal Networking Council Advisory Committee of nonfederal scientists and network users has been created to support the FNC.

The IITA Task Group, originally led by NSF, was chartered to acquire information on IITA-related activities planned and in progress within the HPCC agencies, to develop a plan for coordinating IITA activities among those agencies, to assist HPCCIT in reporting requirements related to IITA, and to provide advice on technology and policy related to IITA to HPCCIT and its parent committees. This group continues under the current leadership of ARPA.

The Task Force on Working Groups is evaluating the current structure and coverage of HPCCIT Working Groups and will make suggestions for changes in modes of operation and coverage.

3.3. Selecting Projects for Funding

To reduce redundancy and encourage adequate funding of all HPCC areas, the HPCCIT Subcommittee has worked with the agencies to define complementary areas of responsibility. Within each agency, the actual mechanisms used to evaluate and fund activities differ. Several agencies conduct R&D by agency staff (and/or national laboratories), while others work through communities of researchers in academia and industry. Each agency that funds external R&D activities has a program to review and fund competitive, merit-based awards that are consistent with agency mission. Calls for proposals for these grants receive wide distribution, including both electronic means (e.g., electronic bulletin boards and Mosaic servers on Internet) and traditional hard copy (e.g., Commerce Business Daily).

3.4. Progress Measures

The primary method of measuring progress of the Program is the establishment and subsequent review of yearly milestones. The FY 1994 milestones for the overall Program can be found in Section 4 of this document. These milestones are firm expectations, based on funding that has been estimated for this fiscal year. The FY 1995 milestones (also shown in Section 4) describe tentative expectations for the Program, based on the anticipated funding at the President's requested level. Achievement of these high level milestones depends on specific program activities that take place within each agency. Section 5 gives a more detailed look at each program activity, including their expected milestones. The combined successful completion of these activity-specific milestones across all agencies translates into meeting the various HPCC Program milestones.

The nature of these milestones (and hence the definition of "measuring progress") depends on the activity undertaken. In some cases, clear quantifiable measures accurately convey progress (e.g., the number of colleges and universities connected to NREN, the speed of network communications, and the computational time needed to solve a specific application problem). In these cases, the HPCC Program and Project Activity milestones use these quantifiable measures as the primary measure of progress. In other cases, a number of different quantifiable measures have been proposed, but their use does not always accurately reflect progress. For example, developing a new computer system with a peak execution speed of 100 gigaflops may not be progress if that rate can only be achieved on artificially constructed test cases. In these cases of uncertain quantifiable measures, a critical aspect of the HPCC research activity is to identify and define useful measures. Finally, in cases when quantifiable measures are unavailable or not necessarily reflective of the progress achieved, the HPCC Program relies on qualitative measures. The HPCC Program combines the advice and recommendations of technical reviewers and advisory committees with program managers' reviews to assess progress.

3.5. Reporting Results and Interactions

The HPCC Program uses a variety of vehicles to convey its results. At the individual projects level, investigators present their results in workshops, conferences, and journal publications. In addition, investigators rely on electronic means (such as the HPCC Software Exchange and file transfer services) to quickly share and distribute their tools. At the agency level, periodic reports summarize the results of the HPCC Program[7]. In addition, the HPCC agencies are beginning to establish electronic information servers to provide direct access to the results of their research projects. At the HPCC Program level, the National Coordination Office publishes a yearly report that summarizes the highlights of the overall Program. The NCO has also established its own information server that is being linked to all of the agency-level servers. When completed, anyone with access to the Internet will be able to access technical results throughout the HPCC Program.

Outreach conferences and workshops play an important role in reporting results of the HPCC Program and receiving feedback on progress. *High Performance Computing and Communications: Technology for the National Information Infrastructure* [8] identifies six HPCC-sponsored conferences and workshops as well as 13 other ones in which HPCC played a prominent role. Additional events (sponsored and/or involving HPCC) since that report include:

- Workshop on Enabling Technologies for Peta(FL)OPS Computing Systems (February 1994),
- R&D for the NII: Technical Challenges (February 1994), and
- ARPA HPCC Symposium (March 1994).

In addition to conferences and workshops, representatives from the NCO and HPCC agencies meet with and provide briefings to companies, universities, associations, media, and other groups. In 1993, the NCO met with over 80 such groups.

---

[7] For example, "The DOE Program in HPCC", March 1993, DOE/ER-0536 and "NASA High Performance Computing and Communications: Level One Plan", August 1993.

[8] February 1994, pages 18-19.

3.6. Program Oversight

The High Performance Computing Act of 1991 (Public Law. 102-194) requires the President to select a High Performance Computing Advisory Committee of nonfederal members, including representatives of the research, education, and library communities, network providers, and industry, who are specially qualified to provide advice and information on high performance computing. While this committee has not yet been appointed, representatives of industrial companies, organizations, and associations have met with the HPCCIT in public sessions to advise the agencies. Detailed program reviews are also carried out by individual agency via their management review structures, including official advisory committees. The NREN especially has benefited by advice of the Federal Networking Council, which is appointed by the Director of NSF. In addition, both CIC and Congress have oversight responsibility for the Program. This section highlights oversight and review activities that have been carried out recently and describes existing review mechanisms within each of the agencies.

The HPCCIT Subcommittee encourages broad comment and external recommendations on the HPCC Program. For example, the Subcommittee has devoted portions of its meetings to hosting visitors. Three recent meetings included representatives from computer system vendors, the telecommunications industry, and software industry.

- The November 1993 meeting included 17 representatives from computer system vendors and consortia. One important message that came out of this meeting was the recommendation to push for more involvement of independent software vendors to get critical applications moved to the emerging massively parallel platforms.
- The January 1994 meeting included 15 representatives from the telecommunications industry and related forums. Several key messages recurred throughout this meeting: (1) a desire by a broad range of companies to participate in providing the information superhighway to the public, (2) the key role of the federal government in working with industry to establish telecommunication standards, and (3) the need for a level playing field (in reference to government regulations) to provide new capabilities to customers.
- The February 1994 meeting included 18 software industry representatives. This meeting produced a much wider spectrum of opinions and issues than the previous two. Commonly mentioned points included: (1) the common base of software development tools must be improved, (2) third party software vendors are reluctant to enter the HPCC market due to its small size and the relatively high development costs, (3) better protection is needed for software as intellectual property and corporate assets, and (4) National Challenge applications that are close to market needs and generate substantial market interest should be stressed.

Similar HPCCIT Subcommittee meetings are in the planning stages to acquire feedback from academia. To further the feedback process from these meetings, the HPCCIT Subcommittee will actively seek participants' comments on this Implementation Plan.

A variety of reviews, studies, and hearings on various aspects of the HPCC Program have taken place in the past 15 months and others are in the planning stages. They include:

- Computer Science and Technology Board of the National Academy of Science commissioned study of the federal HPCC Program (due Summer 1995)
- General Accounting Office study of the ARPA, NSF, DOE, and NASA HPCC Programs. (in progress)

- The NSF report to the Senate Appropriations Committee entitled "Overview of the Management of the NSF HPCC Program" (March 1994)
- Informal briefing for Committee on Information and Communication (March 1994)
- Congressional Staff briefing (February 1994)
- Congressional Staff briefing (September 1993)
- The National Science Board commissioned Blue Ribbon Panel Report on the NSF HPC program[9]
- Testimony before Senate Committee on Commerce, Science and Transportation, Science Subcommittee (August 1993)
- Congressional Staff briefing (July 1993)
- Congressional Budget Office study of the HPCC Program[10]
- Testimony before House Committee on Science, Space and Technology, Subcommittee on Science (May 1993)
- A General Accounting Office Review of ARPA's HPCC Program[11]
- Testimony before House Committee on Science, Space and Technology, Subcommittee on Technology, Environment, and Aviation (March 1993)
- Testimony before House Committee on Energy and Finance, Subcommittee on Telecommunications and Finance (February 1993)
- Congressional staff briefing (February 1993)

The ARPA reviews its HPCC efforts at many different levels to insure consistent program evaluation in a dynamic R&D environment. Projects produce reports on a regular basis that are reviewed by program managers. An annual process of updating accomplishments milestones and project plans is tied to the incremental funding process. Site visits, project meetings, principle investigator meetings, and regular interactions using the Internet as an information infrastructure enable ARPA staff to keep up with their program management responsibilities. In addition, contracting details are handled by contracting agents that work with ARPA as part of the program management process. Office directors and program managers develop plans and milestones that are approved by senior management during the planning and budget cycle. New programs and ideas are proposed during this process. In conjunction with yearly funding decisions by ARPA and DoD, senior ARPA technical management critically review program areas, plans and accomplishments. Guidance is provided to reflect programmatic, technical and funding directions. At the DoD level, programs are described through a formal process that requires Agency, ARPA Comptroller, DoD Comptroller, senior ARPA management, and senior DoD approvals. Once approved, these descriptions become part of the Defense budget submitted to Congress for approval. In addition to other internal Federal HPCC reviews, there are Congressional Briefings, High Performance Computing Agency crosscuts, technical working groups, DoD advisory panels, and a number of National Academy of Science studies which strongly contribute to the planning process.

---

9 "From Desktop to Teraflop: Exploiting the U.S. Lead in High Performance Computing", August 1993.

10 "Promoting High-Performance Computing and Communications", A CBO Study, June 1993.

11 "High Performance Computing: Advanced Research Projects Agency Should Do More to Foster Program Goals", GAO Report to the Chairman, Committee on Armed Services, House of Representatives, GAO/IMTEC-93-24, May 1993.

Numerous processes and mechanisms are in place that contribute to the definition and review of the NSF HPCC program long term goals and objectives. In addition, more specific goals and objectives are established for each of the activities within the program. The National Science Board, panels and committees commissioned to study and recommend program activities, and external advisory committees contribute primarily to the more general, long-term goals. The establishment of goals, objectives, implementation mechanisms, evaluative measures and the broader value of specific activities is accomplished through such means as: external peer review (mail reviews, panel reviews, and site visits), workshops for developing research agendas, program committees of visitors, technical oversight teams, ongoing site visits by program staff and outside experts, program officer review of final project reports, and the bodies of opinion held by the community of researchers themselves. Evaluation of progress is integrated into the management process as appropriate to each class of HPCC activities. Examples are:

- Blue Ribbon Panel for the NSF HPC program
- Federal Networking Council Advisory Committee for parts of the NSFNET
- Directorate Advisory Committees
- Committee of Visitors for each research program

The DOE HPCC program focuses on basic mathematics and computational research and on developing and delivering technology for use by other scientists and engineers in DOE and US industry. Performance evaluation is an integral part of these programs. Because of this focus, external review by prospective users of the technology is a critical component of measuring performance. In previous years this has been most explicitly present in the use of committees of users such as the ESnet Steering Committee and the ER Supercomputer Users Group to evaluate the effectiveness of the access and networking programs. Many of the education programs established under the DOE HPCC program have evaluation built into them. In the area of software technology, using prospective users of technologies as reviewers has ensured that the technology developed is desired by users. In FY 1995 and beyond, DOE will further formalize these procedures to include program-wide reviews of the basic technology components of the program by significant prospective users of those technologies. These will include reviews of effectiveness in all categories as well as specific numerical targets.

The NASA HPCC program evaluations take place at several levels. At the agency level, the NASA Advisory Council (NAC) has established the Ad Hoc Task Force on Supercomputing, which has recently completed a review and report on the NASA HPCC program. It is expected that other such bodies will be permanently chartered under the Aeronautics Advisory Committee to advise the NASA HPCC program. Within the program, comprehensive reviews are conducted for each of the projects annually. In addition to appropriate NASA personnel, representatives from other federal agencies, academia, and industry may be invited to participate. Annual reviews of program progress and plans will also be conducted by the NASA HPCC Working Group. In addition, quarterly reviews of the NASA HPCC program are conducted by program managers, the associate administrator of aeronautics and the director of the NASA HPCC Office.

The NIH HPCC program goals are enhancements of existing NIH program missions to support biomedical science and expand biomedical knowledge. Program objectives are developed by Institute Directors, advisory bodies and senior program staff, and are peer reviewed for determination of merit. Each of the participating NIH components has one or more standing external advisory committees which review new and existing programs. These include the National Library of Medicine (NLM) Board of Regents, NLM Board of Scientific Counselors, National Center for Research Resources (NCRR) Research Resources Advisory Council, NCRR Biomedical Research Technology Review Committee, the National Cancer Advisory Board, the

National Cancer Institute Division of Cancer Biology and Diagnosis Board of Scientific Counselors, and the Division of Computer Research and Technology Advisory Council. The final decision regarding individual HPCC programs within each of the participating ICDs rests with the Director of that Institute. Within each of the participating NIH Institutes, mechanisms exist to ensure objective evaluation of progress and results and identification of remaining obstacles.

The National Security Agency reviews its HPCC support efforts on a yearly basis in several separate reviews. A steering group comprised of senior managers from the technical components provides high level guidance prior to the formal budget process. The steering group receives individual project assessments from the project managers and determines whether any major shifts or changes are needed to the NSA program. Senior management has the flexibility to sponsor HPCC efforts in several budget reviews. Individual projects are proposed and budgeted within the technical components. They are constantly being evaluated by the project managers. Monthly status reports are evaluated and meetings held with the project staff to ensure that the correct focus is maintained. During the budget review cycle, the projects are evaluated and retained based on their performance, importance relative to other initiatives, and priority based on the steering group guidance. New projects can be proposed by the technical components each year during the NSA Technology Program review process.

Annually NIST has the National Academy of Science review all the programs and activities of each operating unit, as part of its normal operation. Assessment includes relevance, performance measures and achievements. Consequently, programs and activities that are a part of the federal HPCC Program are subject to review and comments of an external panel of experts from academia and industry. A written report is presented to the Director of NIST, the Administration and Congress. Selection of individual projects and subsequent progress reviews are conducted by the program manager. These are reviewed by an inter-operating unit panel comprised of senior managers for relevance to agency mission, the program, and for continued acceptable performance.

The NOAA HPCC program supports and enhances NOAA programs in environmental prediction and stewardship. Comprehensive NOAA science reviews of these programs are held periodically, with a review cycle completed as of April 1994. Quarterly reviews of NOAA HPCC program progress are conducted as an integral part of NOAA-wide quarterly reviews by the NOAA Administrator and other NOAA senior line and program managers. Overall NOAA HPCC program goals and plans are reviewed annually as part of the NOAA strategic planning process.

The EPA HPCC program is focused on incorporating advances in computing and communications technology into critical environmental assessment applications and transferring those advanced tools to key State, Federal, and industrial users. EPA senior management review the EPA HPCC program annually to assess its relevance to the agency mission and program achievements. Agency guidelines require an external peer review of the EPA HPCC program every two years. The external review panel is composed of representatives from other federal agencies, academia, and industry. The first EPA HPCC external review is scheduled for summer 1994. Within the program, each major project is reviewed at least twice a year to evaluate progress toward the program objectives.

3.7 HPCC Planning Beyond FY 96

Computing and communications are *ongoing*, strategic enabling technologies. The current HPCC Program provides additional stimulation to accelerate the progress of these technologies and the benefits that accrue from their use. This Program was initiated as a five year effort, covering FY 1992 through FY 1996. Since in-depth planning for the FY 1997 budget will begin in the Spring of 1995, preparations for that planning must begin now. Two avenues of activity need to be pursued: a complete external review of the HPCC Program to ascertain its likelihood of meeting and/or exceeding its goals and an assessment of options and benefits that could accrue from continued efforts to stimulate the technology. Given the recent request to the Computer Science and Technology Board to perform an HPCC Program review due next year, this activity would presumably address a key part of the first avenue. In addition, the HPCCIT Subcommittee will continue to seek input and comment from industry, academia, and other segments of the government at its meetings. To further the feedback process from these meetings, the HPCCIT Subcommittee will actively seek participants' comments on this Implementation Plan.

The assessment of options and benefits that could accrue from continued efforts to stimulate computing and communications technology has already begun. HPCC will continue to sponsor a variety of workshops with technical experts, engage discussions with potential users of the technologies, and prioritize critical directions for future activities. Three HPCC workshops are a part of that process:

- Enabling Technologies for Peta(FL)OPS Computing Systems (February 1994) sponsored by NASA, NSF, NSA, Ballistic Missile Defense Organization, ARPA, and DOE;
- ARPA HPCC Symposium (March 1994); and
- System Software Tools for High Performance Computing Environments (scheduled for September 1994).

The key questions that will be addressed throughout the planning process include:

- What technical objectives in computing and communications show the greatest promise for benefiting the public and enabling U.S. industrial competitiveness with additional stimulation?
- If successful, what results would accrue and who would benefit?
- What are the expected costs and risks associated with the objectives?
- Which agencies are most qualified to contribute toward these objectives?
- What is the right mix of industry/academia/government participation?

Through broad input and extensive discussions, the HPCCIT Subcommittee expects to develop a plan and detailed proposals that will merge naturally into the budgetary planning process.

Table 1: **HPCC FY 1995 IP Budget Summary**

Agency	Budget (BA, $ M)				HPCC Components by 1995 Pres. Req.				
	FY 93 Actual	FY 94 Pres.	FY 94 Est.	FY 95 Pres.	HPCS	NREN	ASTA	IITA	BRHR
ARPA	254.50	343.00	298.90	357.40	110.70	61.10	29.60	140.80	15.20
NSF	225.17	341.00	267.00	328.62	21.70	52.86	141.23	50.64	62.19
DOE	99.51	123.60	122.20	125.40	10.90	16.80	75.50	1.20	21.00
NASA	81.60	121.50	113.00	124.90	9.70	12.70	81.20	17.50	3.80
NIH	47.20	71.50	57.80	81.80	4.90	8.40	23.80	29.10	15.60
NSA		42.23	40.23	40.13	16.10	11.80	11.80	0.23	0.20
NIST	2.10	26.10	18.10	56.40	6.75	3.65	4.60	41.40	
NOAA	9.80	12.40	10.80	25.25		8.70	16.05	0.50	
EPA	8.01	11.90	7.90	14.67		0.70	11.70	0.30	1.97
Totals	**727.89**	**1093.23**	**935.93**	**1154.57**	**180.75**	**176.71**	**395.48**	**281.67**	**119.96**

4. Program Components

This section presents an overall HPCC Program view of the Implementation Plan for each of the five components. The discussion begins by identifying the agency-level funding assumptions on which these plans are based. Each HPCC component discussion provides the following information:

- A brief description of the types of activities included,
- A list of Milestones that have been set for FY 1994 and FY 1995,
- A table of the agency activities supporting the component, and
- A short status report reflecting on progress to date and potential issues for the future.

4.1. Budget Planning Assumptions by Component

Table 1 provides the basic financial planning information used by each of the HPCC agencies in preparing this Implementation Plan. The "FY 94 Pres." and "FY 95 Pres." columns refer to the amounts requested by the President in his budgetary request to Congress each year. The "FY 94 Est." column is the amount each agency has been authorized to spend on HPCC as a result of Congressional action. The last five columns break down the FY 1995 Presidential Request into the planned spending level for each of the five HPCC components. The breakouts by component are estimates, in that many activities span a variety of components and categorization of efforts includes subjective judgment.

For this plan, the FY 1994 Milestones assume the FY 1994 Estimate funding level for each agency. The FY 1995 Milestones assume the FY 1995 Presidential Request funding level in each agency. Likewise, all discussions of program status assume these numbers.

4.2. HPCS -- High Performance Computing Systems

4.2.1. HPCS Definition

Technological leadership in computing and the solution of critical application problems requires continuous enhancement of the computational capabilities of computing systems. New generations of systems must solve previously computationally intractable problems, improve resolution for complex problems, and reduce the time to solution for time-critical problems. The HPCS program addresses both of these objectives, using large computational science problems (Grand Challenges) and large information-based problems (National Challenges) as the primary application targets. The Implementation Plan for HPCS focuses on producing scalable parallel computing systems through the development and evaluation of prototype systems.

This plan contains several critical aspects. *Scalable* computing systems permit incremental expansion of hardware resources to meet the needs of increased application problems. They also permit the use of the same design in smaller scale workstation-size systems. The focus on computing *systems* reflects the need to develop a balance among all of the components that impact the solution of application problems, including the processors, memory hierarchies, interconnection networks, I/O subsystems, and the interplay of hardware and software. The program will develop innovative systems that will provide a 100 to 1,000 fold increase in sustained computing capability compared to computing systems that were in conventional use at the start of the program in FY 1992.

HPCS focuses on technological challenges in the earliest stages of the product development cycle. Critical underlying technologies are developed in prototype form along with associated design tools. Early evaluation of systems throughout the design cycle provides critical feedback for the refinement of successive generations. Larger projects that are closer to yielding commercial products are funded on a cost-sharing basis with industry. HPCS contains four elements: Research for Future Generations, System Design Tools, Advanced Prototype Systems, Evaluation of Early Systems.

Table 2 identifies the major program activities that contribute to the HPCS component. ARPA provides most of the HPCC funding and management for designing and building prototype systems as well as for supporting design tools. NSF and NSA carry out related efforts, with a greater focus on various application-specific needs. NSF, DOE, NASA, NIH, and NSA acquire and evaluate early prototype systems. The evaluations generally take the form of implementing important agency applications (or critical computational kernels) on these systems and comparing performance against the expected potential of the system.

4.2.2. HPCS Milestones

Previous HPCC documents have identified two primary milestones for HPCS after 1993:

- Demonstrations of tenth scale teraops systems technologies (FY 1995)
- 1 teraops system ready for deployment (mid 1990's)

The HPCC Program retains these key milestones. Additional milestones are identified which provide clearer intermediate progress toward these goals and which highlight aspects related to acquisition and evaluation of these systems.

FY 1994

- Produce HPC systems that demonstrate faster clock rates than previously available; current performance targets include 100 MHz clocks for general purpose compute nodes and 200 MHz clocks for superpipelined processors
- Demonstrate scalable mass storage systems with system services and I/O channels.
- Demonstrate embedded prototype small-scale teraops systems with special-purpose mass storage and gigabit network interfaces.
- Develop and demonstrate improved tools for coupling system designs to actual fabrication of products.
- Develop and demonstrate design tools for low power and wireless computing systems.
- Explore the viability of new technologies (e.g., optical computing, processors in memory, superconducting crossbars, and wide bandwidth memory) for use in general-purpose and special-purpose systems.
- Acquire and begin evaluation of three moderate-sized systems with a primary focus on medical, aerosciences, and energy applications.

84-115 O - 94 - 6

Agency Total (BA $M)	Agency / Program Activity	FY 95 Pres. (BA $M)
110.70	ARPA	
	Scalable Computing Systems	60.20
	Microsystems	44.50
	Integrated Command and Control Technology	6.00
21.70	NSF	
	Computing Systems and Components	20.70
	Engineering (non-NC/GC)	1.00
10.90	DOE	
	Advanced Prototype Systems	1.00
	Evaluation of Early Systems	9.90
9.70	NASA	
	Testbeds	9.70
4.90	NIH	
	DCRT High Performance Biomedical Computing Program	4.00
	NCI Frederick Biomedical Supercomputing Center	0.90
16.10	NSA	
	Supercomputing Research	14.10
	Superconducting Research	2.00
6.75	NIST	
	NREN Deployment and Performance Measures for Gigabit Nets & MPP Sys.	0.30
	Metrology for Future Generations of Microelectronics	6.45
180.75	HPCS FY 1995 Total	180.75

Table 2: HPCS Program Activity Summary for FY 1995 Presidential Budget

FY 1995

- Demonstrate tenth scale teraops systems technologies and teraops-class modules with the potential for being components of next generation systems.
- Develop embedded systems capable of higher performance using smaller amounts of space; current performance target is 50-100 gigaflops/cu. ft.
- Continue to explore the viability of new technologies (e.g., optical computing, superconducting electronics, and wireless computing) for use in general-purpose and special-purpose systems.

- Improve design and fabrication tools (e.g., virtual prototyping, advanced packing, and process optimizations).
- Upgrade two systems to larger computational power for medical and energy applications.

4.2.3. HPCS Status

The HPCS component continues to make substantial progress toward meeting the above goals and milestones. The one previously identified FY 1993 milestone, 100 gigaops systems ready for deployment, was achieved on schedule. This milestone was met as part of joint projects with Cray Research, Intel, Thinking Machines, Kendall Square Research, and IBM. New concepts in secondary storage systems have been developed which are revolutionizing the U.S. disk industry and have increased potential for finding information in large, unstructured databases. Technologies for advanced design and packaging tools have also been recently demonstrated. Acquisition of prototype systems (some of which are precursors to later upgrades) includes: Intel Delta and Paragon, Kendall Square KSR-1, and the Thinking Machines CM-5. Results of large applications running on these machines, in some cases, have already produced new results and new capabilities.

Looking forward, the HPCC Program is cautiously optimistic that the milestones identified above and the overall goal of teraops performance can be achieved on schedule, assuming no major changes in the anticipated funding profile. Representatives of industry meeting with the HPCCIT management appeared to confirm this view of the overall goal.

However, the HPCC Program is concerned about the slowed progress in Evaluation of Early Systems, due to funding limitations and priorities. Early acquisition of many small prototype systems by the mission agencies for quick testing and early feedback is underfunded. Agencies have adjusted by buying fewer, moderately-sized "starter systems." They begin testing and evaluation on these systems, with the intent of upgrading them to larger sizes in out years (assuming that the initial evaluation results are favorable). While generally reducing costs, this approach also reduces the number of different prototype systems acquired by mission agencies. This choice reduces opportunities for cross-platform comparisons of important agency applications. ARPA and NSF will continue to deploy prototype systems for early evaluation and testing of some applications.

4.3. NREN -- National Research and Education Network

4.3.1. NREN Definition

The principal objectives of the NREN[12] component are to:

- Establish and encourage wide use of gigabit networks by the research and education communities to access high performance computing systems, research facilities, electronic information resources, and libraries.
- Develop advanced high performance networking technologies and accelerate their deployment and evaluation in research and education environments.
- Stimulate the wide availability at reasonable cost of advanced network products and services from the private sector for the general research and education communities.

---

12 "NREN" is a service mark of the U.S. Government, registered with the U.S. Patent and Trademark Office.

- Catalyze the rapid deployment of a high speed general purpose digital communications infrastructure for the Nation.

The underlying strategy of the HPCC Program has been to support the solution of important scientific and technical problems of broad national significance in collaboration with all interested sectors in government, industry and universities. In the networking component of the program this strategy has led to a unique collaboration in both the research and operational aspects of the NREN activity.

The program provides fundamental and comprehensive extensions beyond existing capabilities of current hardware and software. The goal is to acquire the most advanced networking services that support the ever-increasing networking demands of high performance computing to the extent feasible, while assuring a stable and consistent level of services for its users. The routine purchase of generally available off-the-shelf technology is not a goal of the program, although off-the-shelf technology may be integrated with advanced technology.

The NREN component has two elements -- Interagency Internet and Gigabit Research and Development. Both elements encourage and foster participation of the private sector to maximize the leverage of federal funds. Although the NREN is a research and educational network program, a deliberate consequence of including substantial private sector activity is that the technology and services developed, and even the facilities themselves, may be the model for a more ubiquitous network offering developed under private, or other public efforts. Many industrial research organizations and commercial establishments supporting the nation's scholarly enterprise are connected. In fact, the commercial network segment is the fastest growing segment of the Internet. Federal operation and ownership of network facilities and services, already minimal, are intended to be reduced even further as the program develops. Nevertheless, HPCC Program priorities remain the central focus of the NREN component. While other federal and private sector participants are encouraged, the degree of their participation must be contingent on several factors, such as program focus, cost sharing, and technology leverage.

Currently, and at the end stage of this development, the Interagency Internet program activity will result in a comprehensive service offering to the nation's community of researchers and educators at all levels. The network will connect them to each other and to the facilities and other resources they use in their scholarly endeavor, such as databases and libraries, laboratories, scientific instruments, and computation centers. As a facilitator and enabler of intellectual activity, the Interagency Internet system will allow connectivity to supporting organizations such as publishers and hardware and software vendors. International connections that serve the national interest are also included.

Important features of the Interagency Internet NREN element are:

- Use of telephone company facilities, and not the laying of fiber optic cables or building of a physical network; and
- Driving technology and broadly seeding the market, while avoiding competition with the private sector.

Because of this latter aspect, success of this part of the program inevitably leads to tension and concerns that government services not remain in place once a technology offering has been demonstrated and seeded. The policy of the NREN component is to accelerate this transition to the private sector, while not compromising the need for stable and consistent services by the research and education communities.

Table 3 identifies the major program activities that contribute to the NREN component. NSF coordinates the broad deployment of the Interagency Internet element in support of the HPCC Program. NIH, NOAA, and EPA activities within NREN focus on expanding connectivity. NSF, DOE, and NASA continue dual roles of expanding connectivity and development of future generations of faster and more robust networks. ARPA and NIST also contribute extensively in this latter role of future generations of networks. NSA concentrates primarily on enhancing secure communications over these networks.

4.3.2. The NREN and the NII

Because it is anticipated that the NREN will be the testbed for many NII applications and services, the NII has a significant impact on the direction of the NREN component. This impact increases the importance of alternative bitway physical models (e.g., mobile systems, adaptive bandwidth systems, and systems with asymmetrical bandwidth) to support a variety of connectivity requirements. In addition, this impact increases the importance of developing an infrastructure of network services, such as security and authentication, with standard interfaces, which can be used to build Information Infrastructure Services available uniformly to government and non-government users of the network. It is of critical importance to the NII that it not have to duplicate such services. This requirement increases the importance of standard service interfaces that can be implemented over a variety of physical bitways and is transparent to the applications that require its services.

4.3.3. NREN Status

The goals of the NREN may be grouped into:

- Improved network connectivity,
- Improved network capabilities, and
- Improved manageability and controllability of the network.

Improved network connectivity and improved network capabilities are both critical to the success of the NREN. The usefulness of the network derives from the number of interesting resources (libraries, experiments, computational resources, other people) which are available on it and from the range of ways in which the network enables one to use those resources (electronic mail, distributed computing, desktop video-conferencing, computer mediated collaboration). However, improved manageability and controllability of the network is also critical to the success of the NREN and to its usability as a prototype for the NII bitways. Improved connectivity and capabilities are useless unless the network can be managed and controlled to ensure stable operation. This must include a realistic, scalable, deployable national and international security architecture. Increased commercial operation of the network or of network components, especially in a multi-service provider environment, demands improved manageability and controllability.

In terms of network connectivity, the NREN component is well ahead of schedule, at least for services at T1 and lower speeds. The Interagency Internet forms the core of a nation- and world-wide infrastructure whose size in terms of number of computers attached is growing at over ten percent per month. However, this growth poses important challenges for manageability and controllability. The current IP addressing infrastructure is now predicted to be able to support this growth for about seven years, provided Classless InterDomain Routing (CIDR) continues to be deployed.

Agency Total (BA $M)	Agency / Program Activity	FY 95 Pres. (BA $M)
61.10	ARPA	
	Networking	43.10
	Global Grid Communications	18.00
52.86	NSF	
	NSFNET	40.86
	Very High Speed Networks and Optical Systems	3.00
	Education and Training	9.00
16.80	DOE	
	Energy Sciences Network (ESnet)	14.80
	Gigabit Research and Development	2.00
12.70	NASA	
	NREN	12.70
8.40	NIH	
	NLM Medical Connections Program	6.50
	DCRT High Performance Biomedical Computing Program	1.00
	NCI Frederick Biomedical Supercomputing Center	0.40
	NCI High speed networking and distributed conferencing	0.50
11.80	NSA	
	Very High Speed Networking	3.50
	Secure Operating System Development	5.70
	High Speed Data Protection Electronics	2.60
3.65	NIST	
	NREN Deployment and Performance Measures for Gigabit Nets & MPP Sys.	1.90
	Metrology to support mobile and fixed based communications networks	1.75
8.70	NOAA	
	Networking Connectivity	8.70
0.70	EPA	
	State Network Connectivity	0.70
176.71	NREN FY 1995 Total	176.71

Table 3: NREN Program Activity Summary for FY 1995 Presidential Budget

In terms of increased capability, there have been significant advances in network speed and throughput. Work on very high speed networks appears to be proceeding on schedule as is work on all optical networks and advanced network components. Equally important, network capable applications and tools have been developed (e.g., Mosaic, Gopher, desktop video-conferencing, whiteboards, wide area collaborative environments, and image transport) which have dramatically expanded the usefulness and the capabilities of the network. It is to a significant degree because of these services that connections to the Internet are growing so fast. There has also been significant progress, on a technical level at least, in such areas as authentication and security, distributed file systems, and privacy enhanced mail. The ongoing and planned activities in these areas will permit substantially greater trust of the network by users. Intra-gigabit testbed connectivity has slipped one year due to technical problems with the local telecommunications companies as well as a slow start in terms of some advanced technologies. Finally, there has been some progress, both technical and administrative in establishing information directory services to guide users, although this is still a formidable problem.

In terms of increased manageability, there has been some progress; however, significant challenges remain. Each advance in the network in the other two areas complicates this third area. Rapid expansion in the number of hosts attached to the net requires re-engineering the addressing and routing architecture of the net in the near future as well as ensuring that the net remains manageable and stable throughout the transition. New services such as desktop video require enhanced multicast services from the net which must be designed and managed. Even in authentication and security the administrative issues of key management and choice (either dynamic or static) of algorithms for digital signature and encryption must be settled. It is a significant weakness of the current infrastructure that it requires an irreducible core of "network wizards" to manage the routing between the providers to avoid disaster (e.g., major network portions isolated from the rest of the network). In addition, a deployable architecture to support realistic, manageable network usage accounting and statistics gathering does not currently exist.

4.3.4. NREN Milestones

Previous HPCC documents have identified a key milestone for NREN after 1995:

- By 1996, gigabit research will lead to an experimental nationwide network able to deliver speeds up to 2.4 billion bits per second to individual end user applications.

Even though intra-gigabit testbed connectivity has slipped one year, the HPCC Program expects to achieve this milestone on time. Other NREN milestones follow.

FY 1994

- Develop policy mechanisms, methodologies, and associated software and hardware for ensuring network integrity, controlling access and protecting data from unauthorized use.
- Complete T3 Level 3 and Level 2 NREN interconnects to NASA HPCC Centers.
- Begin ESnet upgrades to 45 Mb/s ATM services at 20 sites.
- Demonstrate satellite-based gigabit applications utilizing ACTS satellite and associated ground terminals.
- Demonstrate advanced visualization and modeling applications on high speed networks.
- Make awards for Regional Network Providers, NAP Manager(s), Routing Authority, and vBNS.

FY 1995

- Design and prototype 10 Gb/s switches for asynchronous transfer mode.
- Develop and implement advanced protocols for high speed information transfer over long distances.
- Make KERBEROS-based authentication services available in full production basis across ESnet.
- Provide production video and collaborative work - telework support across ESnet to support access to DOE experimental facilities.
- Demonstrate techniques for resource reservation guarantees on high performance networks.
- Evaluate gigabit testbed project.
- Deploy prototype backbone Internet Protocol (IP) network at 155 Mb/s.
- Resolve issues of synchronization control and algorithms related to distributed network high performance computing applied to Grand Challenge class problems.

4.4. ASTA -- Advanced Software Technology and Algorithms

4.4.1. ASTA Definition

The computing systems that will be developed by the HPCC Program are not simply faster versions of current computer designs. They represent new architectures for which fundamentally new algorithms and software development will be required. Dramatic improvements in algorithm design and software technology are essential to achieve sustained teraflops computing system performance. Scalable parallel and distributed systems provide new opportunities and challenges to developers of software and algorithms. The Advanced Software Technology and Algorithms component of the HPCC Program has three broad objectives:

- Enable significant progress toward the solution of Grand Challenge application problems in science and engineering;
- Improve system user-friendliness, reliability, software productivity and effectiveness; and
- Use experience from leading edge applications to help guide new software efforts.

These broad objectives are the framework for more specific ASTA objectives adopted by the agencies and used as a basis for the design and conduct of their specific HPCC programs. The ASTA component has four key elements, each with a set of objectives .

- Support Grand Challenges
  - Support a wide variety of computational applications in fields such as: biological sciences, geosciences, engineering, computational mathematics, physical sciences, aeronautics, and energy research.
  - Demonstrate large-scale, multidisciplinary computational research (e.g., accurate molecular structure predictions by multidisciplinary scientific teams) on heterogeneous, distributed systems using the Interagency Internet to access distributed file systems and national software libraries.

- Project the performance of full Grand Challenge applications on teraflops computing systems through controlled testbeds of prototype environments.

- Software Components and Tools
  - Establish portable, scalable libraries that enable accelerated transition from one generation to the next and across different system structures, and improve overall portability of applications code and establish software reuse libraries.
  - Support the evolution of parallel operating systems and the development of a suite of software tools that enhance productivity and effectiveness of both man and machine (e.g., tools for load-balancing, data partitioning, compilers, run-time optimizers, performance monitoring, visualization, and virtual environments).
  - Reduce the entry barrier to high performance computing by providing scalable software technologies and environments in desktop class computing systems.
- Computational Techniques
  - Support for improvement and parallelization of algorithms and data structures.
- High Performance Computing Research Centers (HPCRCs)
  - Increase access to HPCRCs in research communities.
  - Acquire advanced prototype and early production HPCS for use and evaluation.
  - Provide HPCS, links to NREN, computer operation and facilities support and related resources for enabling Grand Challenge applications and as support for the other elements of ASTA.

Table 4 lists agency activities that contribute to the ASTA component of the HPCC Program.

4.4.2. ASTA Status

The ASTA component has aggressively pursued development of large scientific applications and the acquisition of HPC systems on which to run them. Currently, 34 Grand Challenge scientific teams are developing applications on high performance computer systems. In addition, numerous other application codes are being developed and/or ported to HPC systems. Probably the most significant ASTA result to date is the dramatic increase in the number of research areas that are beginning to develop new applications for HPC systems. These applications all need substantial computing resources. At present, HPCC runs over a dozen high performance computer centers, serving over 12,000 users.[13] However, recent estimates indicate that such resources will need to expand further to meet existing needs for the applications currently under development (not to mention new IITA needs).

As expected, the reports of progress vary depending on the complexity of the algorithms and the disparity between the natural data layout in the code and the type of machine organization present in the target architecture. The most successful applications to date tend to contain extremely high amounts of data parallelism that map quickly and easily onto the selected architecture. New software tools are getting into users' hands, including performance monitoring and instrumentation toolkits, 3D visualization tools, message passing libraries, and a parallel linear algebra library. These tools help to make the task of application development easier and

---

[13] *High Performance Computing and Communications: Technology for the National Information Infrastructure,* Supplement to the President's Fiscal Year 1995 Budget, April, 1994, pg. 13.

Agency Total (BA $M)	Agency / Program Activity	FY 95 Pres. (BA $M)
29.60	ARPA	
	Scalable Software	29.60
141.23	NSF	
	Supercomputer Centers	76.43
	Research Centers	6.00
	Grand Challenge Applications Groups	10.75
	Software Systems and Algorithms	16.35
	Human-Machine Interaction & Information Access	6.80
	Biological Sciences (non-NC/GC)	11.50
	Engineering (non-NC/GC)	3.23
	Geosciences (non-NC/GC)	3.84
	Computational Mathematics (non-NC/GC)	1.89
	Physical Sciences (non-NC/GC)	3.77
	Social, Behavioral & Economic Sciences (non-NC/GC)	0.67
75.50	DOE	
	Enabling Energy Grand Challenges	9.00
	Software Components and Tools	12.60
	Computational Techniques	3.40
	Supercomputer Access	35.60
	HPC Research Centers	12.90
	HPC Research Centers-Global Climate Collaboration	2.00
81.20	NASA	
	Testbeds	16.70
	Grand Challenge Support	55.30
	Systems Software	9.20
23.80	NIH	
	NLM Biotechnology Informatics	4.80
	NLM Electronic Imaging	2.20
	NCRR Advanced Software Technology and Algorithms	8.80
	DCRT High Performance Biomedical Computing Program	3.00
	NCI Frederick Biomedical Supercomputing Center	4.60
	NCI High speed networking and distributed conferencing	0.40

Table 4: ASTA Program Activity Summary for FY 1995 Presidential Budget

Agency Total (BA $M)	Agency / Program Activity	FY 95 Pres. (BA $M)
11.80	NSA	
	Supercomputing Research	11.80
4.60	NIST	
	Development and Dissemination of Scientific Software for HPCS	4.60
16.05	NOAA	
	Advanced Computation	16.05
11.70	EPA	
	Environmental Modeling	6.45
	Computational Techniques	5.25
395.48	ASTA FY 1995 Total	395.48

Table 4: ASTA Program Activity Summary for FY 1995 Presidential Budget

more reliable. However, automated tools for partitioning and mapping programs onto HPC systems still need substantially more effort.

The following accomplishments were achieved in FY 1993:

- Released initial version of Mosaic hypertext network application software for a dramatically new way to traverse and collect information on the Internet. Over 100,000 copies distributed in first 8 months.
- Released High Performance Fortran and developed other monitoring tools for HPC environments.
- Deployed 100-gigaflops system to an HPC Center. Deployed several vendors' first models of scalable high performance systems. Benchmarked 50 gigaflops application. Demonstrated high performance workstation clusters.
- Demonstrated initial Grand Challenge applications with notable success in several areas, including: a model of a beating heart, 3D simulation of flutter in an aircraft wing, earthquake ground motion, computational studies in fundamental physics, global circulation models, and other areas.
- Assisted in software conversion and improved and optimized parallel Grand Challenge applications code.
- Established 9 new Laboratories for research in software and protocols to support high performance applications.
- Established a minimally populated HPCC software exchange library for promoting reuse of software.

Looking ahead, two issues will need more attention. First, HPCC will need to take steps to get greater involvement of independent software vendors (ISV) in the Program. ISV's represent a critical sector of the computing community that provides user-friendly applications platforms for public and industrial use. Their applications need to exist on HPC systems. Second, HPCC will need to push for greater success in the area of automated and integrated software tool platforms.

4.4.3. ASTA Milestones

FY 1994

- Demonstrate scalable software libraries, toolkits and prototype HPC programming environments for Fortran and C++. Evaluate and extend to improve support for Grand Challenge applications.
- Continue to develop and improve on efficiency of parallel computational techniques to maximize use of HPC systems.
- Develop infrastructure to support networked digital libraries.
- Provide continued support to Grand Challenge applications through code conversions, optimization and facilities support. Upgrade some HPC systems to the latest model available. Review computational tools for promotion of industrial usage.
- Evaluate the scalability of parallel operating systems, I/O and message passing schemes for real user applications.

FY 1995

- Demonstrate additional system support tools for real-time HPC systems, including monitoring of biomedical experiments and data transfer from telescope to remote data archive.
- Continue development of flexible, interoperable tools for portable parallel program creation, testing, and performance monitoring. Expand definition of High Performance Fortran to handle adaptive, dynamic computations.
- Develop tools for collaborative Interagency Internet research environment.
- Continue support for Grand Challenge and National Challenge applications, including ecological simulations, molecular dynamics, advanced manufacturing, air quality modeling, climate modeling, atmospheric chemistry, digital libraries, combustion, computational chemistry, gene sequencing, and protein structure determination.
- Demonstrate prototype system for access and retrieval of reusable software, with automated expert advice.
- Upgrade supercomputer centers and testbeds with more powerful parallel computing systems to support Grand Challenge and National Challenge applications.

4.5. IITA -- Information Infrastructure and Technology Applications

4.5.1. IITA Definition

Information Infrastructure Technology and Applications (IITA) activities will demonstrate feasible solutions to problems of national importance, such as health care or 21st century manufacturing, using the full potential of the nation's rapidly evolving high performance communications and information processing capabilities. The HPCC Program will produce enabling technologies critical to developing the National Information Infrastructure (NII). With the incorporation of IITA activities, the HPCC Program will advance intelligent system interfaces, real environments augmented with virtual environments, image understanding, language and speech understanding, intelligent agents for augmenting human capabilities, and next generation data and object bases for electronic libraries and commerce. This will be coupled with a vigorous program of testbed experimentation that will ensure the continued leadership of the United States in critical information processing technologies.

IITA will strengthen the HPCC technology base, broaden the markets for these technologies, and accelerate industry development of the NII. Federal HPCC agencies will work closely with industry and academia in pursuit of these objectives. These objectives are to be accomplished, in part, by accelerating the development of readily-accessible, widely-used, large-scale applications with significant economic and social benefit. The HPCC Program's original focus of enhancing the Nation's computing and communications capabilities is thus extended to address a broader set of technologies and applications that have an immediate and direct impact on critical information capabilities affecting every individual in the Nation.

The development of such applications will be predicated on (1) creating the underlying scalable computing technologies for advanced communication services over diverse bitways, effective partitioning of applications across elements of the infrastructure, and other applications support services that can adapt to the capabilities of the available infrastructure; and (2) creating and inserting an intelligent service layer that will significantly broaden the base of computer information providers, developers, and consumers while reducing the existing barriers to accessing, developing, and using advanced computer services and applications. In parallel with these activities, a more effective software development paradigm and technology base will be developed. This will be founded on the principles of composition rather than construction, solid architectures rather than ad hoc styles, and more direct user involvement in all stages of the software life cycle. The entire technology base developed in this program, including services and software, will be leveraged across the National Challenges, leading to significant economies of scale in the development costs.

The IITA program component consist of the following four elements: Information Infrastructure Services, System Development and Support Environments, Intelligent Interfaces, and National Challenges. Table 5 shows a list of the agency activities that contribute toward IITA.

4.5.2. IITA Status

Although the IITA program component was first established in 1994, it builds upon diverse efforts already well underway in the HPCC agencies. For example, ARPA and NASA had already begun to explore high speed satellite-based extensions to the gigabit testbeds in 1993. Mosaic, developed under NSF support at the National Center for Supercomputer Applications (NCSA) at the University of Illinois, provided a network-aware global hypermedia interface in 1993, and has swept the HPCC research community (and beyond) in the last twelve months. ARPA, NASA, and NSF have already been funding virtual reality and telepresence research projects like the CAVE developed at the University of Illinois at Chicago. A number of critical National Challenge applications efforts were also already underway at the HPCC agencies: crisis management at ARPA; advanced manufacturing at ARPA, NIST, NASA, and NSF; environmental monitoring at NASA, NOAA, and EPA; health care at NIH and NASA; and energy-demand management at DOE. Several of the NSF/ARPA-sponsored gigabit testbeds are coupled to National Challenge applications, such as the North Carolina VistaNet testbed, which is being used to support radiation treatment planning in collaboration with NIH.

ARPA, NASA, and NSF have initiated a research program on digital libraries for FY 1994 through FY 1998. The program will incorporate new knowledge and technologies, testbeds to demonstrate results of the research, and university/industry/user partnerships.

Agency Total (BA $M)	Agency / Program Activity	FY 95 Pres. (BA $M)
140.80	ARPA	
	National-scale Information Enterprises	33.90
	Foundations	4.00
	Health Information Infrastructure	9.80
	Information Sciences	15.80
	Intelligent Systems and Software	72.30
	Global Grid Communications	5.00
50.64	NSF	
	NSFNET	5.30
	Research Centers	0.15
	Research Infrastructure	2.30
	Human-Machine Interaction & Information Access	0.60
	Computational Mathematics (non-NC/GC)	0.70
	Physical Sciences (non-NC/GC)	4.00
	Social, Behavioral & Economic Sciences (non-NC/GC)	2.34
	IITA Program	35.25
1.20	DOE	
	Information Infraservices	1.20
17.50	NASA	
	Information Infrastructure Technology	6.80
	Information Infrastructure Applications	10.70
29.10	NIH	
	NLM Intelligent Agent DB searching	4.70
	NLM HPCC Health Care Applications	11.00
	NCRR Information Infrastructure Technology Applications	8.80
	DCRT High Performance Biomedical Computing Program	3.00
	NCI High speed networking and distributed conferencing	1.00
	High Perf. Communications for PDQ, CancerNet, and Electronic Publishing	0.60
0.23	NSA	
	Technology Based Training	0.23
41.40	NIST	
	Development and Dissemination of Scientific Software for HPCS	3.00

Table 5: IITA Program Activity Summary for FY 1995 Presidential Budget

Agency Total (BA $M)	Agency / Program Activity	FY 95 Pres. (BA $M)
	Support for Electronic Commerce	2.75
	Electronic Libraries and Distributed Multimedia Applications	1.25
	Systems Integration for Manufacturing Applications	25.20
	Assurance, Reliability and Integrity of NREN Objects	1.20
	Specification and Testing of High Integrity, Distributed Systems	4.00
	Language, image and text processing	4.00
0.50	NOAA	
	Information Dissemination Pilots	0.50
0.30	EPA	
	Public Data Access	0.30
281.67	IITA FY 1995 Total	281.67

Table 5: IITA Program Activity Summary for FY 1995 Presidential Budget

The Information Infrastructure Technology and Applications Task Group met during the second half of 1993 to identify the technologies that contribute to information infrastructure, to develop a common terminology across the participating agencies, and to inventory on-going and planned projects within these agencies that are associated with IITA. A report describes the work of the committee and its recommendations.[14]

4.5.3. IITA Milestones

FY 1994

- Initiate major research programs in digital libraries, virtual prototyping, problem solving environments, health care information infrastructure, electronic commerce/ procurement, and advanced manufacturing.
- Demonstrate advanced authoring tools and document interchange mechanisms for creating materials suitable for delivering government (and other) information to the public. Demonstrate a national biomedical information kiosk.
- Develop the underlying infrastructure services for authentication, authorization, privacy, and security in order to improve the trustworthiness of the network infrastructure.
- Develop protocols and interfaces for network-based collaborative work environments. Develop advanced computer interfaces in support of human-computer interaction and planning applications.
- Begin to create domain specific digital libraries, large scale databases/data sets, and user interfaces for focused research communities in the social sciences, mathematical sciences, earth sciences, and behavioral sciences.

14 "Information Infrastructure Technology and Applications", National Coordination Office for HPCC, February 1994.

FY 1995

- Extend the research programs begun in 1994 to include reliable computer controlled systems, high assurance systems, telecommuting/telework, infrastructure for environmental research, network infrastructure for education.
- Demonstrate dissemination of real-time geophysical, satellite imagery, and other remote sensed data over the Internet. Develop distributed geographical information systems for the NII. Provide wide access to government statistics databases for researchers and the general public. Investigate how massive information sources such as these can be more effectively utilized, for example, by using advanced search engines and intelligent agent processing.
- Enable domain specialists, such as physicists and engineering researchers, to acquire and distribute information easily and rapidly in the form of electronic preprints accessible over the Internet.
- Establish information infrastructure interoperability testbeds to insure a robust, extensible, and open architecture for the National Information Infrastructure.

4.6. BRHR -- Basic Research and Human Resources

4.6.1. BRHR Definition

The BRHR component is designed to encourage investigator initiated, long term research on computational and experimental projects to maintain the flow of innovative ideas and talented people into high performance computing and communications areas. This component also addresses long term national needs for more skilled personnel, enhancement of education, training, materials and curriculum development in HPCC areas. The BRHR component contains four elements: (1) Basic Research, (2) Research Participation and Training, (3) Infrastructure, and (4) Education, Training, and Curriculum. Table 6 identifies the major program activities that contribute to the BRHR component.

While the primary goal of basic research is to generate new knowledge, it is simultaneously possible to increase the odds that the knowledge will result, ultimately, in useful products to a broader scientific and non-scientific constituency and to expedite the transformation of new ideas into new technologies. This can be achieved by steering the directions of inquiry into promising areas and assigning increased priorities and resources to research projects addressing recognized needs. The HPCC Program places particular value on this dimension of basic research. At the same time, the Program will continue to nourish those larger processes and capabilities which underlie successful basic research generally. These include research infrastructure and tools, communications technologies, mechanisms and models for enhanced multidisciplinary collaboration, and the development of new conduits and occasions for technology transfer between sectors.

4.6.2. BRHR Status

BRHR continues to produce critical research results along a broad spectrum of topics (e.g., modeling molecular dynamics, RNA structure prediction, coherent laser sources, correctness proofs for VLSI processor design, parallel numerical algorithms, and virtual reality). BRHR has expanded educational opportunities and training for all academic levels (e.g., SuperQuest, EarthVision, the National High School Honors Program, Adventures in Supercomputing, graduate assistantships, and post-doctoral research grants). Funding in this category has also helped expand the research infrastructure for parallel computing, high speed networks, and

Agency Total (BA $M)	Agency / Program Activity	FY 95 Pres. (BA $M)
15.20	ARPA	
	Foundations	10.00
	Information Sciences	5.20
62.19	NSF	
	Research Centers	4.40
	Research Infrastructure	18.65
	Software Systems and Algorithms	9.00
	Very High Speed Networks and Optical Systems	8.30
	Human-Machine Interaction & Information Access	3.60
	Computational Mathematics (non-NC/GC)	5.00
	Physical Sciences (non-NC/GC)	2.00
	Education and Training	11.24
21.00	DOE	
	Research Participation and Training	2.00
	Education, Training and Curriculum	3.00
	Basic Research for Applied Mathematics Research	16.00
3.80	NASA	
	BRHR	3.80
15.60	NIH	
	NLM HPCC Training Grants	3.60
	NLM IAIMS grants	5.40
	NCRR Basic Research and Human Resources	5.00
	NCI Frederick Biomedical Supercomputing Center	0.80
	NCI High speed networking and distributed conferencing	0.10
	National Coordination Office	0.70
0.20	NSA	
	Supercomputing Research	0.20
1.97	EPA	
	Education/Training	1.97
119.96	BRHR FY 1995 Total	119.96

Table 6: BRHR Program Activity Summary for FY 1995 Presidential Budget

virtual reality. In addition, efforts to define and implement a cross-disciplinary computational science curriculum are beginning to show acceptance at a number of major universities.

Since the beginning of the HPCC Program, there has been building a stronger call that strategic research principles and practices be adopted for a larger fraction of sponsored basic research programs. For the BRHR component this means focusing investments and efforts more closely around national goals deemed important to economic growth and improvements in the quality of the everyday life. It means that for individual research projects, efforts should be made in advance to identify goals to be achieved, to establish performance benchmarks and progress milestones, to develop appropriate evaluation methodologies and to expend additional effort to ensure that research results be carried beyond publication in scholarly journals to application to practical problems. Each participating HPCC agency has been asked to implement these steps by integrating these principles and practices into their HPCC management approach.

4.6.3. BRHR Milestones

FY 1994

- Expand research infrastructure for virtual manufacturing, networked digital libraries, distributed multimedia systems, and parallel I/O.
- Develop models and tools for reasoning about systems that are reactive, triggered by asynchronous events.
- Continue algorithm and software development for application-specific domains, including DNA sequencing, molecular dynamics, and fluid flow.
- Continue research on computer vision and robotics, with an emphasis on new HPCC applications of these technologies.
- Continue support for educational outreach programs (e.g., SuperQuest, EarthVision, the National High School Honors Program, Adventures in Supercomputing, graduate assistantships, and post-doctoral research grants).

FY 1995

- Develop initial toolkits for interactive, dialogue-based human/computer interaction.
- Continue research infrastructure support for virtual manufacturing, networked digital libraries, distributed multimedia systems, and parallel I/O.
- Demonstrate software to support transparent computations on a heterogeneous collection of geographically distributed computing systems.
- Initiate a research thrust in Virtual Environments to enhance user-based software development.
- Continue and expand algorithm and software development for application-specific domains, including molecular sequence analysis, molecular dynamics, the design of enzyme inhibitors, air quality monitoring, and fluid flow.
- Continue support for educational outreach programs (e.g., SuperQuest, EarthVision, the National High School Honors Program, Adventures in Supercomputing, graduate assistantships, and post-doctoral research grants).
- Expand the program to link and retrieve conceptually related biomedical information from disparate sources within different medical centers.

5. Agency Plans

Implementation of the HPCC Program takes place within each of the HPCC agencies, where the final decisions are made on the funding, management, and evaluation of program activities. This section organizes HPCC activities by agency[15] to better reflect the operational structure of the funding. The level of information does not reflect individual projects. Since many grants are at the level of $100-300 K, a full breakout of each project in a roughly $1 B program would be extremely difficult to assemble and maintain for this plan. Instead, agencies group together projects that are technically related toward reaching similar goals and objectives. Each grouping constitutes a "program activity", most of which range in size from $0.5 M to $40 M. While this structure does track funds clearly, it does possess several drawbacks. Interagency collaborations are often difficult to recognize and may in fact appear as duplication of effort since similar efforts will be repeated within different agency activities. Also, while the HPCC Program broadly describes its activities under the heading of five major components, actual program activities often span multiple components, thus making some of the actual spending difficult to categorize. In fact, since program activities need not follow strict guidelines of spending within each component, better decisions can be made at lower levels of the program implementation as to the most effective use of the funds to meet Program objectives.

Each agency contribution contains three parts:

- Brief overview of the agency's perspective on HPCC efforts,
- Table summarizing financial data and collaborative ties for its program activities, and
- One-page form for each program activity that describes it.

The individual program activity form contains some fields that require explanation. The "Budget Account Code" field is an internal label used by some agencies to track HPCC funding. The budget fields provide actual spending, estimated spending (based on Congressional action), and Presidential requests (Pres.) for various fiscal years. The numbers below the HPCC component labels estimate the breakdown of the FY 1995 request into the HPCC components. These numbers are estimates, in that many activities span a variety of elements and categorization of efforts includes subjective judgment. The "Agency Ties" section of the form permits two labels. A "Partner" label indicates that the corresponding agency is a funding and/or research partner in this activity. Note that two different agencies may view partnering differently (e.g., developing a scalable system in ARPA would be HPCS, but its use for Grand Challenges could be inside ASTA in other agencies). A "User" label indicates that the corresponding agency needs the results of this program activity for another activity. The large field in the center of the form gives an overview description of each activity. The remaining fields provide highlights of major accomplishments and milestones planned for future years. While these entries highlight significant objectives and results, they should not be viewed as complete descriptions of any activity.

The table at the beginning of each agency's section of the report summarizes the financial data and collaborative ties described above. Again here it is important to recognize that component numbers should be considered as estimates and include subjective judgment. The "Partner/User Agencies" field summarizes the Agency Ties material in each activity; however, due to space limitations, not all agencies may be listed in this field for a given activity.

---

15 This plan covers nine HPCC agencies. The Department of Education is an official HPCC agency; however, its funding and reporting of crosscut activities are done through the Committee on Education and Health Resources.

Advanced Research Projects Agency

ARPA is the lead DoD agency for advanced technology research, and has the leadership responsibility for the HPCC program within DoD. This is based on ARPA's history of technical innovations in computer architecture, integrated circuits, networking, and system software. ARPA's strategy is to focus on developing the underlying technology base for high performance computing and communications while other agencies apply these technologies within the context of their mission-specific application focus.

ARPA's HPCS program focuses on developing scalable high performance technologies (both hardware and software) in which a single element can be used as the basis for a workstation while clustered and arrayed elements yield ever higher performance up to the largest and most powerful systems. While transitioning the technology to reach prototype teraops processing in 1996, foundations are being laid for the next generation of petaops systems and beyond. The NETS program develops both high performance networking and hybrid networking schemes that incorporate a wide diversity of transmission modes, including cable TV and wireless. The ASTA program seeks to unify the diverse models of computation for parallel machines to reduce the complexity of parallel software development. The IITA program focuses on developing a robust information infrastructure services architecture through a strategy of technology development and testbed activities. The BRHR program supports foundational research activities and extends the HPCC community to a more diverse population of researchers and users supporting education, training, and historically black colleges and universities.

Note for the following table:

Budget account codes are accurate for FY 94 and beyond, due to a 1993 consolidation.

Advanced Research Projects Agency

Program Activity	Budget Account Code	Partner/User Agencies	Budget (BA, $ M)				HPCC Components by 1995 Pres. Req.				
			FY 93 Actual	FY 94 Pres.	FY 94 Est.	FY 95 Pres.	HPCS	NREN	ASTA	IITA	BRHR
Scalable Computing Systems	ST-19	All HPCC Agencies	44.90	65.70	51.50	60.20	60.20				
Microsystems	ST-19	NSF, NSA	33.50	55.80	46.40	44.50	44.50				
Scalable Software	ST-19	All HPCC Agencies	36.50	29.40	32.30	29.60			29.60		
National-scale Information Enterprises	ST-19	NSF, NASA, NIST	38.00	28.20	14.00	33.90				33.90	
Networking	ST-19	NSF, DOE, NASA, ...	34.80	41.70	40.20	43.10		43.10			
Foundations	ST-19	NSF	13.90	22.40	12.20	14.00				4.00	10.00
Health Information Infrastructure	MPT-07	NIH			2.00	9.80				9.80	
Information Sciences	CCS-02	NSF	30.20	21.10	21.40	21.00				15.80	5.20
Intelligent Systems and Software	ST-11		22.70	64.00	64.40	72.30				72.30	
Global Grid Communications	EE-45	NSA		8.70	8.50	23.00		18.00		5.00	
Integrated Command and Control Technology	IC-03			6.00	6.00	6.00	6.00				
		Totals:	254.50	343.00	298.90	357.40	110.70	61.10	29.60	140.80	15.20

ARPA	Budget Account Code	Budget (BA, $ M)				HPCC Components by 1995 Pres. Req.				
		FY 93 Actual	FY 94 Pres.	FY 94 Est.	FY 95 Pres.	HPCS	NREN	ASTA	IITA	BRHR
Scalable Computing Systems		44.90	65.70	51.50	60.20	60.20				

Scalable Computing spans the full range from Embedable systems, scalable computing components, and desktop workstations to the largest scale heterogeneous systems. It supports architectural research and experimentation to produce balanced scalable parallel computing system prototypes.

The Prototype Scalable Systems element focuses on developing advanced computing system prototypes, scalable computing concepts, high performance system components, and foundations for high performance computing. Prototype systems will be designed, prototyped, and demonstrated at nominal sustained performances of 100 gigaops. These will support the most significant models of computation including multicomputers, multiprocessors, data parallel, systolic, dataflow, and heterogeneous systems interconnect. Requisite software for efficient utilization of these systems will also be developed.

Future generation computing systems technology projects develop and leverage the underlying component, packaging, and scalable concepts, providing the foundation for future generations. These ensure that the required advanced technologies will be available for the teraops systems. Projects focus on high performance components, application of high density packaging, optimized components and architectures, scalability concepts, and petaops foundations.

The Early Evaluation element gets feedback needed in scalable computing from diverse user communities and problem areas. This is accomplished by evaluating the newest hardware/software systems at the earliest opportunity at sites doing research intended for HPC systems and transferring these to the systems architects and software designers in a timely manner. The evaluations will include architecture, scalability concepts, foundation HPC technology, or applicability to application domains. Performance criteria, results, and metrics will be made widely available to insure an open process of interdisciplinary experience and feedback to the system design and research community.

Agency Ties	
ARPA	
NSF	Partner
DOE	Partner
NASA	Partner
NIH	Partner
NSA	Partner
NIST	Partner
NOAA	Partner
EPA	Partner
ED	

FY 1993 Accomplishments	FY 1994 Major Milestones	FY 1995 Major Milestones	Milestone Changes from Prior IP
Demonstrated 100 Gigaops class systems and modules as part of joint projects including Cray Research (T3D), Intel Corporation (Paragon), Thinking Machines Corporation (CM-5), Kendall Square Research (KSR-1), International Business Machines (SP-1, SP-2). Demonstrated major design concepts for teraops high performance computing systems through experiments with 100 gigaops class systems and other advanced design concepts. Produced rapid feedback for 500 gigaops scale computing systems from early evaluation experiments on 10-50 gigaops scale systems.	Develop foundations for petaops per second and terabits systems. Produce first HPC system compute nodes with 100 MHz clocks. Demonstrate software and hardware compatibility between scalable commercial HPC systems and embedable versions. Demonstrate general purpose scalable mass storage systems and associated system services and input/output channels. Demonstrate embedded prototype small scale teraops systems with special purpose scalable mass storage and gigabit network interfaces.	Demonstrate teraops-class modules covering major models of scalable computing, spanning shared and distributed memory models and fine and coarse grain parallelism, that have the potential for being the foundation for next generation and cost-effective units in computing systems. Develop 50-100 gigaflops/cu.ft. militarized embedded systems leveraging commercial scalable computing components and software. Initiate focused development of scalable I/O capabilities for HPC systems. Develop packaging and engineering concepts for embedable HPC systems capable of 100 gigaops per cubic foot.	

ARPA	Budget Account Code	Budget (BA, $ M)				HPCC Components by 1995 Pres. Req.				
		FY 93 Actual	FY 94 Pres.	FY 94 Est.	FY 95 Pres.	HPCS	NREN	ASTA	IITA	BRHR
Microsystems		33.50	55.80	46.40	44.50	44.50				

The program is responsible for critical aspects of the science and technology which enable the design, development, manufacture, and application of advanced microsystems from basic research through small-scale prototyping. Microsystems refers to single or multiple integrated components which can enable innovative computing systems. A key focus is high performance, densely packaged, low power, scalable components that can be accurately simulated, realized in advanced implementation technologies, and rapidly designed into candidate architectures, considering both hardware and software. Enabling technology includes design tools to manage complexity for scalable and embedable units of replication, techniques to rapidly prototype microsystems, and the applications of computing to electronic design. Experimental microsystems will jointly drive the development of tools and critical components of the next generation of scalable systems. Efforts include tools and techniques supporting architecture experiments, application of packaging approaches to scalable systems, innovative prototyping techniques, information science for manufacturing processes, schemes for interconnecting ensembles of microsystems (e.g., backplanes, interconnection networks, and mobile computing and wireline communications networks).

The Design Technology area overcomes current limitations by supporting higher levels of abstractions exploiting HPC, moving towards multi-chip heterogeneous systems from single-chip systems, and exploiting techniques to reduce the design complexity of high performance technologies. Innovative Prototyping supports rapid prototyping of new systems concepts in advanced implementation technologies, including programmable chips, multi-chip modules, 3-D system infrastructure, and low cost standardized modular building blocks. Experimental microsystems conducts small-scale experiments to include new computer architecture components, wireless systems, packaging experimentation, and embedable systems. Finally, Microsystem Services establishes, develops, evolves, and supports access to critical technology in support of research in HPC systems, enabling a broad research community to effectively and economically leverage new technology.

Agency Ties	
ARPA	
NSF	Partner
DOE	
NASA	
NIH	
NSA	Partner
NIST	
NOAA	
EPA	
ED	

FY 1993 Accomplishments	FY 1994 Major Milestones	FY 1995 Major Milestones	Milestone Changes from Prior IP
Demonstrated advanced design technologies including semiconductor process representations, electromagnetic modeling tools on the HPC base, and synthesis of heterogeneous multiprocessors and testable circuits. Developed and demonstrated low cost rapid prototyping multichip module capabilities. Completed design including packaging and cooling technology in support of embedded Touchstone demonstration.	Develop and demonstrate semiconductor virtual process design coupled to actual fabrication line for real-time process control. Enhance and demonstrate direct support of rapid prototyping of MCM technology. Fabricate operational sub-micron diameter vertical Field Effect Transistor for ultra high density read-only memory. Demonstrate 200 Mhz super-pipelined processor. Develop and demonstrate design tools for low power and wireless computing systems. Demonstrate enhanced fabrication services integrated with library management tools.	Extend network-accessible design and fabrication services to include computational prototyping concepts. Develop early module-level synthesis capabilities. Demonstrate wireless computing design environments with early prototypes of high bandwidth, pico cellular, and wireless access points to the wireline infrastructure. Continue development and standardization of process representations. Demonstrate higher levels of process optimization to include low energy. Initial demonstrations of micro-architectures for advanced packaging and scalable units of replication.	

ARPA	Budget Account Code	Budget (BA, $ M)				HPCC Components by 1995 Pres. Req.				
		FY 93 Actual	FY 94 Pres.	FY 94 Est.	FY 95 Pres.	HPCS	NREN	ASTA	IITA	BRHR
Scalable Software		36.50	29.40	32.30	29.60			29.60		

This program supports the development of the advanced software technologies needed to enable the effective use of scalable high performance computing technologies. The new high performance computing technology requires a new software technology base and systems development process. While this technology builds on and extends the existing technology base, it also needs to exploit the opportunities provided by new technologies to meet the software challenge. The overall system architecture is based on user access through workstations and a variety of servers including the latest large scalable HPC servers connected through computational resources of progressively larger scale and higher performance as software matures and high performance systems are needed for larger problems.

The scalable HPC software technologies enable the same technologies used in the largest and highest performance HPC systems to be available in smaller systems, including workstations and single processor desktop computers. This in turn fosters a lower "barrier of entry" in development and use of new software technologies on the new generation of computing systems. The increased affordability and availability of advanced computing provides new opportunities to apply computing to support the software and systems development process. While high performance computing has become essential to the design and development of the hardware technologies, it is only just beginning to be used to support the software development process itself. The increased use of computing to support the software development process provides a fundamentally new opportunity to further improve the overall problem solving process.

Scalable Software consists of four areas: Systems Software (operating system technology in support of HPCC systems), Scalable Image Processing Systems (software and system technology for advanced image and graphical processing), HPC Environments and Languages (programming technology for HPC systems), and Libraries and Library Technology (common architectures and interfaces to increase software reuse in HPC applications programs).

Agency Ties	
ARPA	
NSF	Partner
DOE	Partner
NASA	Partner
NIH	Partner
NSA	Partner
NIST	Partner
NOAA	Partner
EPA	Partner
ED	

FY 1993 Accomplishments	FY 1994 Major Milestones	FY 1995 Major Milestones	Milestone Changes from Prior IP
Demonstrated prototype scalable microkernel operating systems for systems with thousands of processing nodes that may also include multiple processors, advanced system services, and concepts for their extension to time-constrained systems. Demonstrated prototype transparent replication and end user trusted configuration of wide area file system.	Demonstrate scalable libraries for defense-critical problems, such as computational physics and image processing. Develop and distribute HPC software, documentation, performance measurements, and prototype applications using a wide area file system. Demonstrate distributed ADA on scalable HPC systems. Prototype HPC programming environments for standard languages like C++ and Fortran, while developing new languages like Dataflow and new environments like Visual Programming.	Demonstrate real-time operating system support tools for scalable, distributed HPC systems. Demonstrate software development environments for distributed heterogeneous systems on workstation-based tenth scale teraops systems. Experimentally characterize input/output requirements for large- and small-scale computing systems on scalable parallel systems. Demonstrate prototype integrated HPC programming environment for Fortran and C++. Develop portable, real-time fault tolerant operating system software which is compatible with embedable and commercial scalable HPC systems.	

ARPA	Budget Account Code	Budget (BA, $ M)				HPCC Components by 1995 Pres. Req.				
		FY 93 Actual	FY 94 Pres.	FY 94 Est.	FY 95 Pres.	HPCS	NREN	ASTA	IITA	BRHR
National-scale Information Enterprises		38.00	28.20	14.00	33.90				33.90	

This program develops the underlying computer system's technology to enable applications developers to demonstrate prototype solutions to National Challenge problems. The technology is based on those developed for high performance computing and communications, but exhibiting more extensive capabilities for ubiquitous information access, dynamic discovery of and adaptation to new resources, linkage of objects and their dissemination and distribution over internetworked distributed systems, and trusted system operation on behalf of users and applications. To drive the development of these technologies, the program supports integrated systems technology projects for selected National Challenge applications. In particular, the program will develop the underlying technology base for the fundamental enabling applications of Digital Libraries and Electronic Commerce.

The underlying program elements are (1) software and services and (2) applications demonstrations. The former are service technologies available to applications developers, built upon bitways, that provides a ubiquitously available, network-aware, adaptive interface upon which to construct National Challenge applications. The latter are focused demonstration efforts for fundamentally enabling applications, such as Digital Libraries and Electronic Commerce. The technologies developed in support of these will enable other National Challenge applications that require extensive capabilities for large-scale repositories of network-linked objects, adaptation to the surrounding resource environment, usage metering/accounting/payment mechanisms, and privacy-enhanced, secure, and trusted processing.

Agency Ties	
ARPA	
NSF	Partner
DOE	
NASA	Partner
NIH	
NSA	
NIST	Partner
NOAA	
EPA	
ED	

FY 1993 Accomplishments	FY 1994 Major Milestones	FY 1995 Major Milestones	Milestone Changes from Prior IP
Developed and demonstrated a prototype trusted version of the scalable microkernel operating system TMach. Developmental evaluation for both U.S. trusted computer security criteria and European harmonized criteria.	Demonstrate tools for secure communications, e.g., Privacy Enhanced Mail and secure remote procedure call (RPC). Develop advanced data storage & access tools for file systems with intermittent communications, trusted and secure operations, sophisticated access, and multilevel storage management. Develop prototype software environments for distributed applications. Demonstrate initial national digital library for exchange of technical reports. Initiate an interagency initiative to expand library technology.	Develop unified storage mechanisms for service directories, distributed file sys, and object-oriented DB sys. Prototype common authentication, authorization, accounting, usage metering infrastructure. Prototype approaches for service discovery and interaction. Develop applications building blocks for function partitioning and migration. Develop technology to enable spontaneous commerce on open networks. Develop Electronic Copyright Management System for fully electronic copyright registration, recording, rights transfer, and management.	National-scale Information Enterprises brings together a number of related research threads from among the original four components of the ARPA HPCC program. It represents ARPA's efforts in Information Infrastructure Services and part of the agency's efforts in National Challenge applications.

ARPA — Networking	Budget Account Code	Budget (BA, $ M) FY 93 Actual	FY 94 Pres.	FY 94 Est.	FY 95 Pres.	HPCC Components by 1995 Pres. Req. HPCS	NREN	ASTA	IITA	BRHR
		34.80	41.70	40.20	43.10		43.10			

This program is developing high performance networking technologies as enablers of a world-wide ubiquitous information infrastructure. The driving forces are the need to provide enabling technology in support of global simulation, high resolution remote imagery exploitation, NREN Grand Challenges, and other programs requiring high data rate wide and local area communication and information services. This program will produce networking technologies needed to enable the effective interconnection of the high performance computing systems produced by the HPCS part of the ARPA HPC program. The Networking program is organized into Internet and High Performance Networking.

The Internet area provides the glue between differing types of network "bitways" and different generations of bitway technology. Internet deals with the problems of legacy by providing internetworking paths from the slowest/oldest technology to the fastest/newest and deals with the scaling problems of fielding hundreds or thousands of networks and possibly billions of end systems. Also under this area are the composition of useful user applications such as audio/video teleconferencing from provided and newly developed services and protocols. Finally, the area provides for security services needed for reliable and survivable operation of the networking infrastructure.

High Performance Networking is concerned with increasing the speed of the underlying network technologies as well as innovative ways of delivering bits to the customer. Special issues include satellite, broadcast, telecommunications standards compliance, optical, and affordable local area designs. The objective is to create a suite of complementary systems which can be integrated into an overall system with all of the strengths and none of the weaknesses of the individual systems. Through the use of testbeds, a continually escalating level of performance for application development is provided while validating the effectiveness of the supporting technologies.

Agency Ties	
ARPA	
NSF	Partner
DOE	Partner
NASA	Partner
NIH	Partner
NSA	Partner
NIST	Partner
NOAA	
EPA	
ED	

FY 1993 Accomplishments	FY 1994 Major Milestones	FY 1995 Major Milestones	Milestone Changes from Prior IP
Demonstrated high performance networking in gigabit testbeds and identified limitations for future work. Demonstrated asymmetric networking, coupling dial-up telephone and cable television resources. Demonstrated prototype all-optical network using wave division multiplexing.	Demonstrate Command, Control, and Communications systems technology with scalable high performance network technology enabling full multimedia real-time information exchange using early gigabit networks. Demonstrate prototypes of gigabit SONET/ATM technology operating over fiber and satellite media. Demonstration of all-optical Local Area Network. Demonstrate medical, terrain visualization, and modeling applications on 100 Mbit and Gbit-class networks.	Demonstrate cross-country gigabit and networking technologies coupled with high performance computing capabilities. Demonstrate more advanced Internet capabilities including more diverse bitways, such as cable and wireless links, with embedded intelligence to improve ease of use. Demonstrate techniques for rate adaptive quality of service negotiation in asymmetric networks. Demonstrate bandwidth and service reservation guarantees for networks in support of real-time and critical services. Demo secured routing systems. Perform initial interconnections among gigabit testbeds.	

ARPA	Budget Account Code	Budget (BA, $ M)				HPCC Components by 1995 Pres. Req.				
		FY 93 Actual	FY 94 Pres.	FY 94 Est.	FY 95 Pres.	HPCS	NREN	ASTA	IITA	BRHR
Foundations		13.90	22.40	12.20	14.00				4.00	10.00

The main aspects of the Foundations program consists of (1) Foundations projects that provide a base of information about the rapidly advancing technologies which enable collaboration among different communities to encourage new and more effective interdisciplinary activities; and (2) Community outreach programs that involve broader communities designed to increase the flow of innovative ideas and to provide infrastructure needed to support research and education activities. The Foundations projects fall into the following areas: high performance components, high density packaging scaling concepts, system design tools, foundations of petaops and terabits systems, and a crosscutting of basic research and studies of technology trends, limitations, alternatives, and transitions to broader communities. These programs generally take the form of small to medium size projects that are intended to demonstrate the feasibility of new concepts. Larger projects also include basic research topics that are relevant to a particular project.

The Community outreach program builds the infrastructure necessary to support research and education/training activities in the area of high performance computing. A variety of community services and human resource building programs such as government-sponsored studies, graduate fellowship programs, and other community programs are supported to accelerate technology transition. Community outreach programs seek to expand existing activities and initiate new efforts to improve K-12 education and training opportunities in high performance computing and communications technologies, computational science and engineering for both students and educators. The goal of this program is to change not only the way we educate our children, but to inculcate ourselves with the belief that learning is a lifelong process. This paradigm shift will require the development of a suite of powerful learning tools such as multimedia object bases, system wide assets, courseware and authoring tools, virtual labs and ultimately adaptive teaching techniques.

Agency Ties	
ARPA	
NSF	Partner
DOE	
NASA	
NIH	
NSA	
NIST	
NOAA	
EPA	
ED	

FY 1993 Accomplishments	FY 1994 Major Milestones	FY 1995 Major Milestones	Milestone Changes from Prior IP
Initiated a program of Virtual Reality technology in high performance computing and communications. Initiated research in real-time 3D imaging and multimedia computing. Granted initial awards to four Historically Black Colleges and Universities (HBCUs). Initiated studies in HPCC and NII to foster participation by the research community. Increased education and training activities in HPCC by establishing network based educational testbeds. Initiated a program in postdoctoral and graduate fellowship awards in HPCC.	Initiate program in support of information-intensive applications of HPCC technologies in health care, digital libraries, manufacturing, and education and training. Establish local pilot projects. Demonstrate coast-to-coast video workshops on high performance Internet. Initiate FY94 HBCU participation through BAA solicitation, evaluation, and selective awards. Invite HBCU's to participate in program reviews and promote teaming efforts between schools.	Evaluate Virtual Reality technology program. Visit and provide feedback to the initial four HBCUs that were awarded contracts in FY93. Evaluate research and human resource programs. Review HBCU program to assist in the evaluation of the program. Evaluate infrastructure and support further development. Establish education/training pilot testbeds. Prototype network access experiments.	

ARPA	Budget Account Code	Budget (BA, $ M)				HPCC Components by 1995 Pres. Req.				
		FY 93 Actual	FY 94 Pres.	FY 94 Est.	FY 95 Pres.	HPCS	NREN	ASTA	IITA	BRHR
Health Information Infrastructure				2.00	9.80				9.80	

This program will develop an intelligent, agent-based clinical associate software system and demonstrate its application in Ambulatory Care and Combat Casualty/Trauma Care. The program vision is to greatly improve the tools that support decision-making by clinical users and patients/consumers. The program goals are to: 1) conduct visionary pilot projects of CLASS that demonstrate significant impact in health care, and 2) develop a health care-specific information infrastructure that promotes use/reuse of component-based architectures for health information systems. Models of the health care domain are constructed using a Scenario-based Engineering Process. Domain Specific Software Architecture tools are used to derive a reference architecture, reference requirements and generate instances of user-based evaluation systems (visionary pilots). Important features and attributes of the sub-domain pilots are generalized to produce an associate systems reference architecture for health care (health care specific tools). The key functional components this intelligent systems architecture include (1) Human computer interaction manager, (2) Task/context manager, and (3) Information broker. This architecture stresses multimodal user interfaces, information integration from heterogeneous sources, multi-modal information acquisition, communication and presentation, intelligent on-line support for medical personnel, logistical planning and resource allocation, and simulation-based systems for decision-makers in health care management. It provides an unique environment to develop tools and techniques for integrating and leveraging existing and emerging HPCC technologies. Transfer of the associate system and health care reference architectures into both operational use and the nation's industry base is a key criterion for success. Technology transfer plans include: development of generic technology that migrates to diverse platforms; work in partnership with users, developmental consortia and functional managers; and utilize actual clinical scenarios and operational environment for developing, testing and evaluating; and provide demonstrations with associated qualtitatively improved performance.

Agency Ties	
ARPA	
NSF	
DOE	
NASA	
NIH	Partner
NSA	
NIST	
NOAA	
EPA	
ED	

FY 1993 Accomplishments	FY 1994 Major Milestones	FY 1995 Major Milestones	Milestone Changes from Prior IP
	Develop domain models for ambulatory and trauma/combat casualty care. Demonstrate CLASS - Composite Health Care System mediator. Pilot CLASS components in one clinic at National Naval Medical Center	Develop CLASS reference architecture for ambulatory care. Capture clinical data directly from physicians in two different clinics. Extend CLASS architecture to provide trauma guidelines directly to medics. Demonstrate shared-map planning for distributed health and human services.	None - New Program.

ARPA	Budget Account Code	Budget (BA, $ M)				HPCC Components by 1995 Pres. Req.				
		FY 93 Actual	FY 94 Pres.	FY 94 Est.	FY 95 Pres.	HPCS	NREN	ASTA	IITA	BRHR
Information Sciences		30.20	21.10	21.40	21.00				15.80	5.20

Software technology develops advanced concepts for methods and tools to produce high assurance software, language concepts that facilitate the rapid specification and evolution of systems; and techniques to manage shared complex structured data objects in larger heterogeneous, distributed information systems. Intelligent systems technology focuses on advanced techniques for knowledge representation, reasoning, and machine learning to enable computer understanding of spoken and written language and to advance methods for planning, scheduling, and resource allocation. Human-computer interaction technology focuses on design methods and enabling technology for more natural interaction between people and computers. Microelectronic science calibrates fundamental concepts to produce reliable, testable, and high performance design. High Performance Computing (HPC) science generates concepts and methods for validating and verifying design components, and unique approaches to rapidly develop high performance libraries across multiple HPC architectures.

(Information below continues onto the next page.)

Agency Ties	
ARPA	
NSF	Partner
DOE	
NASA	
NIH	
NSA	
NIST	
NOAA	
EPA	
ED	

FY 1993 Accomplishments	FY 1994 Major Milestones	FY 1995 Major Milestones	Milestone Changes from Prior IP
Initial distribution of a prototype persistent object base using electrical, mechanical, and software design tools.	Develop benchmark problems and metrics for advanced research in information sciences.	Provide Internet access to benchmark problems, metrics, and test data sets and conduct experimental evaluations involving multiple intelligent systems and software technologies.	
Applied a shared object base to concurrent eng. support processes.	Develop advanced concepts for machine learning, automated reasoning, knowledge representation, and advanced information processing for human language understanding & large-scale planning, scheduling, and resource allocation methods.	Develop initial prototype of reusable machine learning, reasoning, and knowledge representation methods and experimentally evaluate advanced information processing methods for human language understanding, planning, scheduling, and allocation.	
Completed proof of correctness for a substantial VLSI processor chip.			
Demonstrated viability of modular interfaces formalism.	Develop design concepts for interactive, dialogue-based human computer interaction		
Demonstrated intelligent mechanisms for information visualization and speech recognition in realistic application domains for interactive problem solving.	Develop process models for prototyping large-scale systems.	Develop initial tool kits for interactive, dialogue-based human computer interaction.	
	Develop advanced concepts for understanding software, high assurance software, and software system composition.	Develop initial language-based methods for software understanding, high assurance, and software system composition.	
Developed and demonstrated improved mechanisms for rapidly constructing and evolving individual software modules.			

ARPA	Budget Account Code	Budget (BA, $ M)				HPCC Components by 1995 Pres. Req.				
		FY 93 Actual	FY 94 Pres.	FY 94 Est.	FY 95 Pres.	HPCS	NREN	ASTA	IITA	BRHR
Information Sciences (continued)										

Agency Ties	
ARPA	
NSF	
DOE	
NASA	
NIH	
NSA	
NIST	
NOAA	
EPA	
ED	

FY 1993 Accomplishments	FY 1994 Major Milestones	FY 1995 Major Milestones	Milestone Changes from Prior IP
Fully automated machine translation using statistical modeling methods with 70 percent comprehension. Developed concepts for software specification language for component-oriented approaches to large-scale software systems, for reuse of systems architecture, and for active vision and vision-guided navigation. Demonstrated integration of object-oriented database and programming language technologies. Developed approaches combining model-based vision with neural networks, and tools that verify adherence to formalized interface specifications; demonstrated approaches on standard protocols.	Develop design concepts of advanced components needed for highly reliable computing systems including mobile, high performance, and graphical systems. Develop advanced concepts for high performance libraries to support multiple parallel architectures and integrated with compiler technology.	Develop initial language-based methods for software understanding, high assurance, and software system composition. Develop initial prototypes for heterogeneous, distributed software system architectures and tools to support construction and maintenance of advanced intelligent systems. Experimentally evaluate library research which supports multiple parallel architectures.	

ARPA	Budget Account Code	Budget (BA, $ M)				HPCC Components by 1995 Pres. Req.				
		FY 93 Actual	FY 94 Pres.	FY 94 Est.	FY 95 Pres.	HPCS	NREN	ASTA	IITA	BRHR
Intelligent Systems and Software		22.70	64.00	64.40	72.30				72.30	

Develop new information processing technology concepts that lead to fundamentally new software and intelligent systems capabilities. This will enable advanced information systems (involving both humans and computers) to more effectively accomplish decision making tasks in stressful, time sensitive situations and create efficient software systems supporting computer and software intensive defense systems. Major areas of technical emphasis are in: (1) intelligent systems (artificial intelligence) including autonomous systems, interactive problem solving and intelligent integration of information from heterogeneous sources; (2) software development technology including languages, algorithms, data and object bases, domain specific software architectures, software prototype technology, software design tools, software reuse, and advanced software engineering environments; and (3) manufacturing automation and design engineering, including the development of advanced software systems that support sharing of engineering knowledge, advanced product and process design representations, integrated product and process design, software tools for design process management, manufacturing process planning, manufacturing process control and demonstrations.

(Information below continues onto the next page.)

Agency Ties	
ARPA	
NSF	
DOE	
NASA	
NIH	
NSA	
NIST	
NOAA	
EPA	
ED	

FY 1993 Accomplishments	FY 1994 Major Milestones	FY 1995 Major Milestones	Milestone Changes from Prior IP
Demonstrated reuse of knowledge among systems, including Integrated Weapons Systems Databases; developed and validated standards of knowledge transmission; demonstrated these in industrial prototypes. Integrated independent software tools to demonstrate integration of data sharing and message passing; demonstrated support for the test and analysis of real time systems and integrated these into a development environment. Demonstrated robust recognition with 12% word error with a 20,000 word vocabulary, in continuous speaker-independent speech. Released beta compiler for image understanding, scalable parallel computer for vision applications.	Develop test cases and testbeds for autonomous systems, command and control, and manufacturing systems. Evaluate the integration of intelligent systems and software in autonomous vehicles. Release the beta version of the Image Understanding Environment and develop advanced methods for vision guided navigation. Develop initial capabilities for human-aided machine translation, document understanding, and speech understanding in adverse conditions. Develop advanced real-time planning and control algorithms. Develop knowledge-based decision aids to support crisis action planning.	Experimentally evaluate the integration of advanced intelligent systems and technologies in multiple autonomous vehicles. Upgrade the Image Understanding Environment and develop prototype implementations of advanced methods for vision guided navigation and cartographic modeling. Develop initial prototype implementations for human-aided machine translation, document understanding, and robust speech understanding in adverse acoustic conditions. Develop initial prototype implementations of advanced real-time planning and control algorithms.	

ARPA	Budget Account Code	Budget (BA, $ M)				HPCC Components by 1995 Pres. Req.				
Intelligent Systems and Software (continued)		FY 93 Actual	FY 94 Pres.	FY 94 Est.	FY 95 Pres.	HPCS	NREN	ASTA	IITA	BRHR

Agency Ties	
ARPA	
NSF	
DOE	
NASA	
NIH	
NSA	
NIST	
NOAA	
EPA	
ED	

FY 1993 Accomplishments	FY 1994 Major Milestones	FY 1995 Major Milestones	Milestone Changes from Prior IP
Incorporated stereo vision into autonomous systems; developed robust systems to aid real time planning of autonomous systems. Deployed configured data sets for research on image understanding, stereo vision, planning, hybrid control, and machine learning. Developed and demonstrated agent based architectures for sharing design knowledge, manufacturing process planning, and manufacturing control. Developed architecture for a heterogeneous databases that integrates and maintains a single image of non-interoperable intelligence community databases. Developed a machine learning strategy to enable computers to learn user interaction characteristics.	Develop advanced methods for information fusion, aggregation, summarization, and explanation. Develop initial language-based methods for describing & composing domain-specific software. Develop initial software environment supporting composition tool integration and software development and testing. Develop evaluation and design concepts to support highly distributed, wide bandwidth info processing applications. Enhance agent based architectures for sharing design knowledge, manufacturing process planning, and manufacturing control.	Enhance knowledge based decision aids to support the rapid construction of multiple crisis action plans. Develop concepts and implementations of scalable machine intelligent algorithms for autonomous associate systems. Develop initial prototype implementations of advanced methods for information fusion, aggregation, summarization, and explanation. Experimentally evaluate language-based methods for describing domain specific software architectures. Develop information infrastructure services for manufacturing, e.g., access to engineering analysis and rapid prototyping.	

ARPA	Budget Account Code	Budget (BA, $ M)				HPCC Components by 1995 Pres. Req.				
		FY 93 Actual	FY 94 Pres.	FY 94 Est.	FY 95 Pres.	HPCS	NREN	ASTA	IITA	BRHR
Global Grid Communications			8.70	8.50	23.00		18.00		5.00	

This program develops and demonstrates advanced communications technologies needed for defense and intelligence operations for the 21st century. The program will develop advanced information processing concepts to support a geographically dispersed staff for crisis management. Services for an enhanced information infrastructure to support command and control will be developed and shown to be applicable to advanced, high performance (and commercially available) networks. This program will demonstrate that commercial communications resources and technologies can be integrated with advanced optical components developed in this program as well as DoD tactical and satellite technology developed elsewhere. The key elements are:

- Applications such as intelligent decision aids, that enable a geographically distributed planning staff to develop and analyze a course of action within hours.
- Advanced services such as scalable file systems, databases, and distributed computing support that are integrated with high performance computing.
- Demonstration networks that validate the R&D and enable early application development and technology transition into DoD efforts.
- Develop network controls pertaining to management and security software technologies to enable end-to-end applications combining all network media.
- Develop advanced optoelectronic network component technology and network architecture for scalable and modular networks. The aggregate network bandwidth will be in the range of terabits per second and the network will handle Multi Media service for both digital and analog signals.

Agency Ties	
ARPA	
NSF	
DOE	
NASA	
NIH	
NSA	Partner
NIST	
NOAA	
EPA	
ED	

FY 1993 Accomplishments	FY 1994 Major Milestones	FY 1995 Major Milestones	Milestone Changes from Prior IP
Program start in FY94.	Initiate network management, control, signaling efforts and demonstrate interoperability between commercial and DoD network assets. Initiate optoelectronic network component technology development: switch, multiplexer, filter, amplifier and synchronizer.	Integrate DoD and commercial networks and demonstrate services and network management in support of "crisis management" applications. Demonstrate advanced optical network capability and demonstrate multi-wavelength reconfigurable network architecture.	

84-115 O - 94 - 7

ARPA	Budget Account Code	Budget (BA, $ M)				HPCC Components by 1995 Pres. Req.				
Integrated Command and Control Technology		FY 93 Actual	FY 94 Pres.	FY 94 Est.	FY 95 Pres.	HPCS	NREN	ASTA	IITA	BRHR
			6.00	6.00	6.00	6.00				

This program develops technology and manufacturing capability for high definition displays and scalable image processing systems. Major components of the program include: projection, head-mounted and direct view displays based on multiple technologies, display architectures and processors, compression algorithms, and high speed data transmission. These efforts will establish a domestic technical capability and demonstrate the manufacturing of components necessary for military systems that capture, process, store, and distribute and display high resolution images.

Agency Ties	
ARPA	
NSF	
DOE	
NASA	
NIH	
NSA	
NIST	
NOAA	
EPA	
ED	

FY 1993 Accomplishments	FY 1994 Major Milestones	FY 1995 Major Milestones	Milestone Changes from Prior IP
	Develop imaging systems and processes needed to realize high information throughput.	Continue developing enabling component technologies for performance and cost goals for high definition image processing systems. Design and fabricate radio-based communications modules and components.	

National Science Foundation

The NSF HPCC program supports and elaborates upon Federal Program goals of extending US technological leadership in high performance computing and communications, accelerating wide dissemination and application of technologies to speed the pace of innovation and to serve the national interests in many critical areas, and spurring gains in US productivity and industrial competitiveness through the use of high performance computing and networking technologies. Program objectives, as the term is used here, refers to more specific ends, the attainment of which signals a major step forward toward achieving programmatic goals. For the NSF HPCC program, the objectives include:

* developing national research and education networking services and capabilities for connecting universities, high schools, research laboratories, libraries, and businesses at speeds of up to one billion bits per second
* introducing new generations of scalable parallel high performance computers and software technologies in order to achieve performance of one trillion computer calculations per second on application areas representing Grand Challenges
* generating fundamental knowledge with the potential for radically changing the state of high performance computing and communications
* creating a cadre of scientists , engineers, and technical personnel knowledgeable in the ideas, methods and value of computational science and engineering and prepared to take advantage of these new capabilities
* encouraging industrial partnerships and affiliations to enhance innovation, technology transfer and U.S. productivity and industrial competitiveness
* making advanced computing and communications information infrastructure available to a larger segment of the society to solve Information intensive National Challenges and advance education

The NSF strategy for meeting its goals and objectives consists of balanced programs of support for individual investigators performing long-term curiosity-driven HPCC research, small group research teams studying single, broader problems, Grand Challenge Applications Groups working on complex problems requiring multidisciplinary teams, Science and Technology Centers, targeted on major and significant research areas, deployment of HPCC infrastructure (general availability of networking services, access to specialized high performance computing capabilities, and provision of local small-scale state-of-art computing instrumentation), and developing new opportunities and technologies for enhancing science and engineering education.

National Science Foundation

Program Activity	Budget Account Code	Partner/User Agencies	Budget (BA, $ M)				HPCC Components by 1995 Pres. Req.				
			FY 93 Actual	FY 94 Pres.	FY 94 Est.	FY 95 Pres.	HPCS	NREN	ASTA	IITA	BRHR
NSFNET		All HPCC Agencies	30.10	43.20	41.16	46.16		40.86		5.30	
Supercomputer Centers		ARPA, NASA, NIH	63.89	80.40	67.76	76.43			76.43		
Research Centers		ARPA, NASA	9.20	9.00	9.78	10.55			6.00	0.15	4.40
Research Infrastructure			18.65	19.20	19.05	20.95				2.30	18.65
Grand Challenge Applications Groups		ARPA, EPA	7.00	13.00	7.75	10.75			10.75		
Computing Systems and Components			18.76	26.90	18.76	20.70	20.70				
Software Systems and Algorithms			21.79	31.50	23.46	25.35			16.35		9.00
Very High Speed Networks and Optical Systems		ARPA, DOE, NASA	9.80	14.40	10.00	11.30		3.00			8.30
Human-Machine Interaction & Information Access		ARPA	10.40	14.68	10.40	11.00			6.80	0.60	3.60
Biological Sciences (non-NC/GC)			7.80	11.00	8.80	11.50			11.50		0.00
Engineering (non-NC/GC)			2.17	5.00	3.10	4.23	1.00		3.23		
Geosciences (non-NC/GC)		NASA, NOAA	1.15	3.25	2.37	3.84			3.84		
Computational Mathematics (non-NC/GC)			5.72	6.22	6.68	7.59			1.89	0.70	5.00
Physical Sciences (non-NC/GC)			2.75	6.35	5.32	9.77			3.77	4.00	2.00
Social, Behavioral & Economic Sciences (non-NC/GC)			0.65	1.00	1.15	3.01			0.67	2.34	
Education and Training		ARPA	15.34	19.90	20.04	20.24		9.00			11.24
IITA Program		ARPA, NASA		36.00	11.42	35.25				35.25	
		Totals:	225.17	341.00	267.00	328.62	21.70	52.86	141.23	50.64	62.19

NSF	Budget Account Code	Budget (BA, $ M)				HPCC Components by 1995 Pres. Req.				
		FY 93 Actual	FY 94 Pres.	FY 94 Est.	FY 95 Pres.	HPCS	NREN	ASTA	IITA	BRHR
NSFNET		30.10	43.20	41.16	46.16		40.86		5.30	

The purpose of the NSFNET Program is to provide for the data networking needs of the U.S. research and education community. The Program operates by supporting operation of the worldwide Internet, and also by funding enhancements to and participation in it.

More than 1100 U.S. colleges and universities have been connected to the Internet through the NSFNET program, and well over a thousand high schools have had their connection facilitated. Libraries, medical schools, and public health facilities have also been connected. The program has directly stimulated the emergence of a vigorous and highly-competitive private-sector industry in Internet hardware, software, and connectivity in which the U.S. is a world leader with an overwhelmingly positive balance of trade.

A new, quasi-operational very high speed network backbone service (the "vBNS") will link NSF-supported high-performance computing centers, and links are being competitively established to scientific applications that demand the performance available. For less demanding, general-purpose networking needs of the community, the Program supplies funds to regional networks - most of whom have their roots in regional university consortia, and, under the Connections Program to individual academic institutions needing Internet connectivity. Together with networking programs in other Federal agencies, the NSFNET program participates in funding administrative functions of the Internet, and collaborates in provisioning international Internet links. The NSFNET program also supports technical development in such areas as database access and bibliographic protocols, routing and addressing, security and privacy, and network management.

Agency Ties	
ARPA	Partner
NSF	
DOE	Partner
NASA	Partner
NIH	Partner
NSA	User
NIST	User
NOAA	User
EPA	User
ED	User

FY 1993 Accomplishments	FY 1994 Major Milestones	FY 1995 Major Milestones	Milestone Changes from Prior IP
Solicitation for Regional Network Providers, NAP manager(s), RA, and vBNS was issued and an adequate number of responsive proposals were received. The Global Schoolhouse project using email and low-tech video conferencing among elementary school pupils in three states and the UK was initiated and an "event" held during National Science and Tech. Week; it generated worldwide enthusiasm and will continue. Three collaborative cooperative agreements made to implement registration, information, and directory services for the Internet; the "InterNIC" awards went to AT&T Network Solutions, Inc., and GA . More than 100 new connections made, including medical and public health institutions, institutions of higher learning, and agricultural extension offices.	Extend Merit award for transitioning to new NSFNET architecture. Make awards for Regional Network Providers, NAP Manager(s), RA, and vBNS. Issue solicitation for additional International Internet Services. Issue solicitation for vBNS connections to qualifying high-bandwidth applications.	Complete transition to new NSFNET architecture. - Oversee Merit's audit of sub-award to ANS. - Close out Merit cooperative agreement for old NSFNET architecture. Begin very high speed Backbone Network Service (vBNS) to NSF supercomputer centers at ≥ 155 mb/s. With other elements of CISE, provide networking support for National Challenge educational activities. Make targeted awards in the network theory initiative. Connect vBNS to at least 3 NAPs at ≥45 mb/s. Support connection of 30 additional institutions to the Internet. Make 2 awards in vBNS Connections Program.	None

NSF	Budget Account Code	Budget (BA, $ M)				HPCC Components by 1995 Pres. Req.				
		FY 93 Actual	FY 94 Pres.	FY 94 Est.	FY 95 Pres.	HPCS	NREN	ASTA	IITA	BRHR
Supercomputer Centers		63.89	80.40	67.76	76.43			76.43		

The four NSF Supercomputer Centers represent major activities which serve the computational needs of all NSF science and engineering disciplines by providing over 8000 users from 49 States access to state-of-the-art high performance computing resources. Additional activities at the Centers include information on and access to emerging technologies; software tools for high performance computing; support for Grand Challenge applications; and education, training, and outreach at all levels. The Centers have pioneered partnerships with the private sector working to introduce HPC technologies into national industries to solve design and manufacturing problems. They are also involved in testing the next generation network capabilities, including developing tools for ease of network navigation. The strategic goals of the Centers are:

Provide access to state of the art equipment to academic community
Research emerging technologies; interact with vendors
Develop tools and software for all aspects of HPC
Lead in computational science applications; Grand Challenges
Provide education and training in the use of High Performance Computing (vector & scalable systems, workstation access)
Act as a resource for State/Local Centers
Educational outreach at all levels

Agency Ties	
ARPA	Partner
NSF	
DOE	
NASA	Partner
NIH	Partner
NSA	
NIST	
NOAA	
EPA	
ED	

FY 1993 Accomplishments	FY 1994 Major Milestones	FY 1995 Major Milestones	Milestone Changes from Prior IP
First Cray C90 and first Cray parallel system (T3D) installed at Pittsburgh Supercomputer Center. Installation of the IBM scalable computer (SP1) at Cornell Theory Center, jointly funded by NSF, ARPA, NIH, and New York State. National distribution Mosaic software by National Center for Super-computing Applications and its acceptance as a platform for network information retrieval. Since software release (May 1993), 100,000 copies distributed. The 1993 Forefronts in Computing Award given to researchers at San Diego Supercomputer Center for determination of structure of protein kinase whose signal is essential for function of many hormones including adrenaline. Active partnerships involving 88 industrial affiliates.	Implementing Andrew File System as cross-center national single user name space file system creating metacenter software infrastructure. Completing initial evaluation phase of scalable parallel computers, jointly with 3 NASA centers. Emphasizing exploration of scalable computing by major segments of the scientists, to create capabilities of engineers, and students to utilize the Centers' resources in production mode. Cooperative efforts with State and Regional Centers to broaden the base of usage of HPCC, and expand education and training activities, with a focus on IITA goals. Establish 4-6 Regional Alliances.	New systems/upgrades of current systems at Supercomputer Centers to more powerful parallel computing systems needed to support academic computing community. Enhancement of metacenter activities especially in the area of mass storage for the national file system being developed at Centers. Additional Regional Alliances (4-6) to enhance computing and communications expertise of individuals in university, education, public and private sectors. Host annual Supercomputing meeting in San Diego, with Center Directors as General Chair and Program Chair (5,000 attendees anticipated).	None

NSF	Budget Account Code	Budget (BA, $ M)				HPCC Components by 1995 Pres. Req.				
		FY 93 Actual	FY 94 Pres.	FY 94 Est.	FY 95 Pres.	HPCS	NREN	ASTA	IITA	BRHR
Research Centers		9.20	9.00	9.78	10.55			6.00	0.15	4.40

The HPCC component provides support for four Science and Technology Centers. These centers share several important characteristics: a unifying cross-disciplinary intellectual focus; an emphasis on knowledge-transfer and linkages with private sector organizations; and significant education and outreach components. These centers focus on

- making parallel computers as easy to use as conventional computers;
- building and displaying models that are visually and measurably indistinguishable from real world entities;
- applying discrete mathematics and theoretical computer science to the solution of fundamental problems in science and engineering;
- human cognition, perception, natural language processing, and the application of parallel computation.

Agency Ties	
ARPA	Partner
NSF	
DOE	
NASA	Partner
NIH	
NSA	
NIST	
NOAA	
EPA	
ED	

FY 1993 Accomplishments	FY 1994 Major Milestones	FY 1995 Major Milestones	Milestone Changes from Prior IP
Release of High Performance Fortran, a specification of extensions to Fortran that can serve as a programming interface to new parallel machines appearing on the market. Collaboration between physicians and the Graphics and Visualization Center resulted in visualization of the beating heart cycle and valve action within the aortic chamber. Solution of the largest scheduling problem ever attempted for the optimized motion in automated manufacturing processes. Summer programs to expose high school teachers and students to the role of computer graphics in high school education including a program for 400 junior high girls. Begin a special year on Massively Parallel Computing.	Evaluate the expressibility of High Performance Fortran using Grand Challenge applications in order to identify areas of emphasis for second round of development. Disseminate and support a Message Passing Interface Standard that will provide vendors with a standard that they can implement efficiently for performing message passing among processors in MIMD distributed-memory computing systems. Begin a special year on Mathematical Support for Molecular Biology at DIMACS focused on the application of discrete mathematics and theoretical computer science on problems in DNA sequencing and in protein structure. Enhance operational Toolkit for construction of 3-D objects with improved interactive capabilities.	Release a set of software tools for the support of experimental analysis in discrete mathematics and theoretical computer science Release a portable prototype library for sparse direct matrix computation that can be optimized for different computing systems. Develop an object-oriented multi-disciplinary analysis and design tool set to solve optimization problems in industrial design. Demonstrate with biological modeling, a fluent, physically based modeling language that extends geometric modeling languages and operators. Release a second version of High Performance Fortran that provides support for adaptive, dynamic computations.	None

NSF	Budget Account Code	Budget (BA, $ M)				HPCC Components by 1995 Pres. Req.				
		FY 93 Actual	FY 94 Pres.	FY 94 Est.	FY 95 Pres.	HPCS	NREN	ASTA	IITA	BRHR
Research Infrastructure		18.65	19.20	19.05	20.95				2.30	18.65

The Research Infrastructure activity aids in the establishment, enhancement, and operation of major experimental facilities and in the acquisition of equipment such as workstations in order to support research activities in the areas of computer and information science, computer engineering, or computational science. In general, support is provided for equipment, maintenance, technical support staff, and other appropriate costs. The groups supported range from single researchers requiring workstations to cross departmental or cross institutional groups requiring access to special purpose instrumentation.

The largest part of the activity is the Institutional Infrastructure Program. The program supports both the acquisition of Scalable Parallel computers for research in parallel computing and computational science and engineering and the acquisition of facilities for experimental research.

The Research Instrumentation program supports experimental research with awards for workstations for experimental facilities.

Agency Ties	
ARPA	
NSF	
DOE	
NASA	
NIH	
NSA	
NIST	
NOAA	
EPA	
ED	

FY 1993 Accomplishments	FY 1994 Major Milestones	FY 1995 Major Milestones	Milestone Changes from Prior IP
Nine new awards for scalable parallel computing were made at a total value of $10.85M. Examples include: Development of a laboratory to explore the CAVE three dimensional virtual reality laboratory; Development of laboratory to explore high speed network protocols; Development of six laboratories incorporating scalable, parallel computers for research in parallel computing. $3.8M was provided for continuing support of 27 major facilities including a massive memory project at U. C. Berkeley, and an advanced virtual manufacturing laboratory at the University of Utah. $3.5M was provided for acquisition of specialized equipment including workstations.	Development of several distributed, heterogeneous facilities to address high speed networking needs of scalable parallel computing research. Infrastructure support for virtual manufacturing research. Development of infrastructure to support networked digital libraries research. Infrastructure support for distributed multi-media systems research. Infrastructure support for research on the input/output problems with scalable parallel systems.	Development of infrastructure to support networked applications in education, advanced manufacturing, and digital libraries. Infrastructure support for Gigabit network research. Infrastructure support for virtual manufacturing. Infrastructure support for multi-media systems research. Continued support of identified infrastructure needs for basic experimental research.	New program for the infrastructure needs of the IITA will be initiated.

NSF	Budget Account Code	Budget (BA, $ M)				HPCC Components by 1995 Pres. Req.				
		FY 93 Actual	FY 94 Pres.	FY 94 Est.	FY 95 Pres.	HPCS	NREN	ASTA	IITA	BRHR
Grand Challenge Applications Groups		7.00	13.00	7.75	10.75			10.75		

This HPCC activity supports multidisciplinary groups of scientists, engineers, mathematicians and computer scientists to apply emerging high performance computing and communications systems to advance the solution of fundamental problems in science and engineering. These activities will generate significant new research in mathematics, computer science, engineering and other scientific disciplines. The groups are listed below.

High Performance Computing for Learning; Massachusetts Institute of Technology
Black Hole Binaries: Coalescence and Gravitational Radiation ; University of Texas
Earthquake Ground Motion Modeling in Large Basins; Carnegie Mellon University
Adaptive Coordination of Predictive models with Experimental Observations; Stanford University
High Performance Computational Methods for Coupled Field Problems and GAFD Turbulence; University of Colorado at Boulder
High Capacity Atomic-Level Simulations for Design of Materials Modeling; California Institute of Technology
Parallel I/O Methodologies for I/O-Intensive Grand Challenge Applications; California Institute of Technology
Massively Parallel Simulation of Large Scale, High Resolution Ecosystem Models; University of Arizona
High Performance Computing for Land Cover Dynamics ; U. Maryland
High Performance Imaging in Biological Research; Carnegie Mellon University
Computational Biomolecular Design; University of Houston
The Formation of Galaxies and Large- Scale Structure; Princeton University
Radio Synthesis Imaging: An HPCC Application; University of Illinois at Champaign-Urbana
A Distributed Computational System for Large Scale Environmental Modeling; Carnegie Mellon University
Understanding Human Joint Mechanics Through Advanced Computational Models; Rensselaer Polytechnic Institute
Advanced Computational Approaches to Biomolecular Modeling and Structure Determination; University of Illinois

Agency Ties	
ARPA	Partner
NSF	
DOE	
NASA	
NIH	
NSA	
NIST	
NOAA	
EPA	User
ED	

FY 1993 Accomplishments	FY 1994 Major Milestones	FY 1995 Major Milestones	Milestone Changes from Prior IP
A cell multipole based molecular dynamics code, including support for automatic load balancing, has been used to simulate a 270K atom model of Teflon on the KSR-1 machine. A 3-D simulation of flutter of an aircraft wing was demonstrated using a PVM- controlled heterogeneous parallel computation paradigm. A beta version of software to support access to the digital library system for unprocessed telescope data has been distributed to the astronomy community. A speech production-based model for a new method of speech recognition that overcomes noise limitations of existing systems has been developed and tested.	Simulate molecular system such as HIV protease using a molecular dynamics code and a Brownian dynamics code. Develop and incorporate compiler enhancements for automatic generation of the equations for sensitivity analysis. Demonstrate an intuitive user interface for an integrated system combining a microscope, CCD camera, controllers and an SGI workstation interfaced to a C-90. Demonstrate language learning hardware on the CM-5. Demonstrate on the CM-5 a version of the molecular dynamics code EGO optimized using the Charm parallel programming system.	Demonstrate the utility of direct sensitivity analysis in the air quality model of the Los Angeles basin. Demonstrate the ability of the AIM microscope system to monitor biomedical experiments and automatically detect events of interest Demonstrate on the CM-5 a simulation of a watershed using a discrete event dynamic model coupled to a geographic information system data base with visualization on a SGI based system Bring the real-time transfer of data from the BIMA telescope system in California to the NCSA data archive to full, routine operational status. Make 5 new Grand Challenge awards.	No new awards in FY 1994.

NSF	Budget Account Code	Budget (BA, $ M)				HPCC Components by 1995 Pres. Req.				
		FY 93 Actual	FY 94 Pres.	FY 94 Est.	FY 95 Pres.	HPCS	NREN	ASTA	IITA	BRHR
Computing Systems and Components		18.76	26.90	18.76	20.70	20.70				

Advances in computing systems and components requires that prototype systems be built quickly and cheaply, and new kinds of design tools need to be developed for this to take place. Prototype systems allow evaluation of new application-driven and technology-driven systems by "exercising" them with real applications which provide realistic tests of novel ideas and performance. Included is research on: systems architecture, including application-specific systems and memory hierarchy; early prototyping and evaluation of hardware and software systems; new tools for systems level automated design and prototyping; distributed design and intelligent manufacturing capabilities; and optical communications systems and devices.

Agency Ties	
ARPA	
NSF	
DOE	
NASA	
NIH	
NSA	
NIST	
NOAA	
EPA	
ED	

FY 1993 Accomplishments	FY 1994 Major Milestones	FY 1995 Major Milestones	Milestone Changes from Prior IP
The Supercomputer Toolkit at MIT used to build a special purpose computer for predicting orbits in the solar system. New concepts in secondary storage systems developed which increased our ability to find information in large, unstructured databases. A software tool, the Rapid Interconnect Circuit Evaluator (RICE) is in use at a number of companies including Texas Instruments, IBM, Intel, LSI Logic, and Motorola reducing evaluation time while providing enhanced accuracy. Workshop held on advanced memories and wide-bandwidth memory and I/O for high performance computers.	The high-performance rapid prototyping facility at the University of Michigan will be made available for use over the Internet, speeding development of new high-performance computing systems. Experiments in measurement and evaluation of new memory architectures will be initiated at testbeds at USC and the University of Virginia. Conduct workshop on directions in optical computing research areas for high-performance computing systems. Develop program in virtual prototyping for manufacturing.	Initiate projects to design and analyze new memory architectures. Initiate projects applying high performance computing to manufacturing, in response to a program announcement on virtual prototyping. Initiate projects in optical computing as recommended by the workshop in FY 1994.	None.

NSF	Budget Account Code	Budget (BA, $ M)				HPCC Components by 1995 Pres. Req.				
		FY 93 Actual	FY 94 Pres.	FY 94 Est.	FY 95 Pres.	HPCS	NREN	ASTA	IITA	BRHR
Software Systems and Algorithms		21.79	31.50	23.46	25.35			16.35		9.00

Research emphasizes the development of scalable parallel algorithms and software technologies, and the development of problem solving environments. Areas of research include: parallel languages and optimizing compilers, parallel operating systems, performance evaluation and prediction, high performance systems for numeric and symbolic computations, parallel algorithms and data structures, algorithms for biological applications, problem solving environments, software engineering, computer graphics, computational geometry, and real-time systems.

In addition, a significant fraction of the activity is devoted to fundamental research, with potential impact on high performance computing, in areas such as the theory of computing, software engineering and theoretical aspects of computer systems and operating systems.

The activity also includes a post doctoral program for interdisciplinary computational scientists.

Agency Ties	
ARPA	
NSF	
DOE	
NASA	
NIH	
NSA	
NIST	
NOAA	
EPA	
ED	

FY 1993 Accomplishments	FY 1994 Major Milestones	FY 1995 Major Milestones	Milestone Changes from Prior IP
Developed software tools for the analysis, partitioning, and allocation of tasks. Developed parallel algorithms for optimization based on linear programming, and adaptive methods for partial differential equations. These algorithms proved effective in new areas such as x-ray analysis. Formulated a new concept for characterizing "strong tractability," a performance property of approximate comp. problems. Computability and computational complexity of quantum (non- Boolean) computation devices were also explored. Supported definition of a common message passing interface for distributed memory parallel systems for portability. Broadened experience for junior computational scientists through the postdoc program which brings them into contact with applications.	Develop efficient parallel algorithms for numeric and symbolic computations with structured matrices. Exploit randomization in parallel algorithm design. Develop models and tools for reasoning about systems that are reactive, triggered by asynchronous events. Apply algorithms based on computational geometry to flexible automation and advanced manufacturing. Develop artificial intelligence interfaces to guide dynamically a solution process by modifying certain characteristics and parameters of a parallel numerical algorithm.	Develop compilers for high performance Fortran for several parallel machines. Develop scalable multiprocessing operating systems. Develop efficient probabilistic algorithms based on information- based complexity for important scientific problems. Demonstrate software engineering techniques for developing and analyzing designs and software with real-time or other critical dependability constraints. Demonstrate software to support transparent computations on a heterogeneous collection of computing systems that are geographically distributed.	None.

NSF	Budget Account Code	Budget (BA, $ M)				HPCC Components by 1995 Pres. Req.				
		FY 93 Actual	FY 94 Pres.	FY 94 Est.	FY 95 Pres.	HPCS	NREN	ASTA	IITA	BRHR
Very High Speed Networks and Optical Systems		9.80	14.40	10.00	11.30		3.00			8.30

The focus of work in this program area is basic research on technology for the transfer of information from one point in space to another (transmission and networking) and/or one point in time to another (information storage and retrieval) including multiple access techniques to enable effective collaborative access to information resources.

Research issues include: the efficient utilization of spectral bandwidth for wireless and wired systems, new architectures and multiple access techniques for lightwave systems, optical technologies, reliable and secure transmission or storage of information in local and in multiple access environments, and fundamental aspects of the characterization, control, and management of information networks.

Areas of currently funded research include: gigabit testbed research, design and analysis of gigabit switching systems, protocols, and software structures for network management, resource discovery among collaborative information spaces in large decentralized environments, network information theory, multisender and multi-receiver network security, modulation, detection, and coding of reliable information storage and retrieval, all optical networks, and optical technologies for computing and communications.

Agency Ties	
ARPA	Partner
NSF	
DOE	User
NASA	User
NIH	
NSA	
NIST	
NOAA	
EPA	
ED	

FY 1993 Accomplishments	FY 1994 Major Milestones	FY 1995 Major Milestones	Milestone Changes from Prior IP
Deployment of high-speed (622 Mb/s, 1.2 Gb/s, 2.4 Gb/s) network facilities. Installation of the first ATM switch in a Central Office (GTE at VISTANET). Development of TENET, a network protocol suite for real-time traffic such as multimedia. Synoptics Communications, Inc. manufactured and delivered commercial 155 Mb/s ATM switch based on switch design from the University of Washington. Open Software Foundation, a not-for- profit technology transfer org.. with industry involvement is evaluating the x-kernel for possible inclusion in future releases of its OS.	Run all Testbed applications on the gigabit testbed facilities. Explore Gigabit Network Problems including characterizations of gigabit network traffic running on operational gigabit testbeds. VISTANET's Radiation Treatment Planning Application evaluated against standard treatment planning modes in use across the country.	Report on the accomplishments of the gigabit testbed project, lessons learned, strengths and weaknesses of technologies deployed in the testbeds. Conference in November to discuss testbed results.	None.

NSF	Budget Account Code	Budget (BA, $ M)				HPCC Components by 1995 Pres. Req.				
		FY 93 Actual	FY 94 Pres.	FY 94 Est.	FY 95 Pres.	HPCS	NREN	ASTA	IITA	BRHR
Human-Machine Interaction & Information Access		10.40	14.68	10.40	11.00			6.80	0.60	3.60

The focus of research in the area of human- machine interaction and information access is in the following areas:

- Scientific Databases: Integration of scientific and engineering methods with well-established database techniques to enhance the design, capabilities and ease of use of the next generation information systems;
- Human Language Technology: Development of technologies for speech recognition, voice input to computers and devices, and machine aided-translation;
- Virtual Environments/Collaboration Technology: Sharable computing and communications environments which many can access, interact across, and use effectively;
- Knowledge-Based Systems: Development of expert systems technologies for use as standard tools for decision making and problems solving in knowledge intensive fields.

Agency Ties	
ARPA	Partner
NSF	
DOE	
NASA	
NIH	
NSA	
NIST	
NOAA	
EPA	
ED	

FY 1993 Accomplishments	FY 1994 Major Milestones	FY 1995 Major Milestones	Milestone Changes from Prior IP
A special initiative "Research on Scientific Databases" launched to promote integration of scientific and engineering methods with database techniques. Workshop held on "Directions Towards an Infrastructure for Temporal Databases" aimed at identifying a common infrastructure foundation for implementation, standardization, and further research on temporal databases. Two collaborative testbeds established, one to support research in space physics and the other in upper atmosphere science. A special research initiative on Human Language Technology launched.	Development of a generic model for multi-media information delivery aimed at the general public. The model will be built on the research results from human computer interaction, virtual environments, and knowledge systems. Completion of the first phase of research for the two testbeds in collaboration technology and demonstration of the methodologies and tools developed in these projects. Continuation of several research projects in computer vision and robotics jointly funded by ARPA to break new scientific grounds and explore new applications of these technologies.	Demonstration of a prototype system capable of collecting, compiling and delivering information in mixed media and modalities: sound, speech, texts, images, graphics. Initiation of a new research thrust in Virtual Environments with the aim of shifting the software development paradigm to a more distributed, user-centered, and domain-oriented focus. Development of new and extended thrusts and directions for the Science and Technology Center on Cognitive Science at U. Penn to provide sharable resources in human language technology over the NREN to the larger research community and the industry.	None.

NSF	Budget Account Code	Budget (BA, $ M)				HPCC Components by 1995 Pres. Req.				
		FY 93 Actual	FY 94 Pres.	FY 94 Est.	FY 95 Pres.	HPCS	NREN	ASTA	IITA	BRHR
Biological Sciences (non-NC/GC)		7.80	11.00	8.80	11.50			11.50		0.00

Many areas of biology - both basic and applied - require the use of computational tools to resolve significant questions. HPCC has had a dramatic impact on many of these areas. Three examples are mentioned below. Each one must analyze experimental results, construct detailed theory and then interrelate the two. They must also develop database systems to handle heterogeneous databases and to develop networking technologies to link various databases. This program activity supports the development of necessary computational tools: software, optimization techniques, and large volume data accessing, storage and managing techniques. In addition, the activity supports modeling of biological phenomena on new computer architectures, which will allow new questions to be posed and answered in these areas. In many cases projects are supported which involve multidisciplinary groups of investigators.

Structural Biology: The challenge is to determine macromolecular structure, and to relate structure to function. Determining structure of a molecule directly from experimental data or inferring it through comparison with known structures are computationally intensive problems that will not be solved without HPCC systems. The impact of solving these problems will be felt in both basic and applied biology, e.g.,rational drug design, as well as biomaterials.

Neurosciences: The challenge is to understand how networks of neurons work, and ultimately how the entire nervous system operates, from sensory systems to learning and memory. New HPCS architectures allow for explorations of these questions. The impact of further understanding of experimental and theoretical systems will be felt in the areas of health care (neuropharmacology, psychotherapy), artificial intelligence and robotics.

Ecology: The challenge is to understand the relationship of individuals that interact at various scales, both spatial - from microorganisms to higher organisms, from populations and to entire ecosystems, ultimately to global scale - and temporal - from nanoseconds to geological. This complexity of the data and of the relationships demands computational solutions. Impact of this area will be felt on issues of global change, biodiversity, and the environment.

Agency Ties	
ARPA	
NSF	
DOE	
NASA	
NIH	
NSA	
NIST	
NOAA	
EPA	
ED	

FY 1993 Accomplishments	FY 1994 Major Milestones	FY 1995 Major Milestones	Milestone Changes from Prior IP
Demonstration of feasibility of linking parallel and serial supercomputers to work on problems in the area of structural biology. Demonstration of remote control of electron microscope via network. Parallelization of a general neural simulation system for computation neurobiology on a heterogeneous network of workstations (as well as massively parallel platforms) and use of the system for semi-automated parameter searchers of complex realistic neuronal simulations.	Development of tools for ecology on parallel architectures, in particular for multi-species models. Development of a center for ecological synthesis and analysis. Development of tools for "on-line" high performance computing (e.g. real-time 3-D imaging of biological specimens during experimentation). Development of algorithms for molecular dynamics with more atoms and longer time steps.	Development of collaborative internet research environments (e.g. Khorus). Expansion the activities of the center for ecological synthesis and analysis, prototyping the linking of heterogeneous databases, expanding the network to more sites. Continued development of computational tools for ecological simulation studies, and distribution of existing tools; likewise for molecular dynamics and neural simulation tools, with aims of increasing size and realism of simulations.	Greater emphasis on information processing (collaborative research environments and "on-line" high performance computing).

NSF	Budget Account Code	Budget (BA, $ M)				HPCC Components by 1995 Pres. Req.				
		FY 93 Actual	FY 94 Pres.	FY 94 Est.	FY 95 Pres.	HPCS	NREN	ASTA	IITA	BRHR
Engineering (non-NC/GC)		2.17	5.00	3.10	4.23	1.00		3.23		

The ENG Directorate supports fundamental engineering research by individual investigators and small groups that contribute to three components of the HPCC initiative: ASTA, BRHR, and IITA.

(1) Software tools and algorithms are developed and applied to the solution of computationally complex engineering problems utilizing parallel and distributed computing environments. Support is also provided for engineering groups addressing grand challenge problems, under NSF's Grand Challenge Application Groups initiative.

(2) Basic research is conducted on optical and optoelectronic technologies that are at the cutting edge for future advances in ultra-high-capacity computing and communications. Emphasis is on integration at the interface of device and system level research that can help accelerate the implementation of these technologies.

(3) Information infrastructure technology and applications issues are addressed in an initiative involving multidisciplinary groups of engineers, computer scientists, and educators to integrate critical information systems on national challenges, such as in health care delivery, civil infrastructure systems, advanced manufacturing, environmental research, and engineering education.

Agency Ties	
ARPA	
NSF	
DOE	
NASA	
NIH	
NSA	
NIST	
NOAA	
EPA	
ED	

FY 1993 Accomplishments	FY 1994 Major Milestones	FY 1995 Major Milestones	Milestone Changes from Prior IP
Development of computational tools and algorithms in the solution of computationally complex engineering problems. Continued support of NSF Grand Challenge Application Groups, in earthquake ground motion, fluid dynamics, human joint mechanics, and large data set storage and analysis. Awards to small research groups in All Optical Networks initiative, jointly with CISE. Optical device research on coherent laser sources, fiber amplifiers, optical switching and modulation.	Begin initiative, with CISE and EHR participation, for multidisciplinary group research on national challenges relating to IITA, including health care delivery, civil infrastructure systems, advanced manufacturing, and engineering education. Expand optical communications systems research directed to high connectivity, multi-gigabit interconnection of remote processors. Seed funding to establish interagency support for US users in US/Japan Joint Optoelectronics Project, to expand the user base in optoelectronics fabrication, and to stimulate growth of a domestic optoelectronics industry.	Continue and expand scope of the FY 1994 ENG initiative on national challenges in IITA, to include other engineering emphases such as environmental research. Continue support of innovative approaches to computationally complex engineering problems utilizing parallel and distributed computing environments. Continue research that bridges the interface between devices and systems in optical and optoelectronic technologies for computing and communications.	None.

NSF	Budget Account Code	Budget (BA, $ M)				HPCC Components by 1995 Pres. Req.				
Geosciences (non-NC/GC)		FY 93 Actual	FY 94 Pres.	FY 94 Est.	FY 95 Pres.	HPCS	NREN	ASTA	IITA	BRHR
		1.15	3.25	2.37	3.84			3.84		

Geosciences and the Office of Polar Programs HPCC activities build on a substantial infrastructure within the directorate. Incremental funds are used to enhance present activities and to synergistically integrate existing efforts into a geosciences program that is responsive and effective in using emerging technologies:

The acquisition and testing of parallel computers of various architectures to determine which technology offers the most promising capabilities. This initial work will guide the acquisition of a massively parallel machine in the FY 95-97 time frame.

Software which effectively uses parallel computers for geoscience problems is being developed and enhanced including parallel algorithms to be used in models such as the Community Climate, global atmospheric chemistry, and eddy-resolving global ocean circulation.

Extended activities to provide access to unique scientific and informational resources by connecting the Antarctic science facilities to the internet for high speed data transmission of global data, images from telescopes and general interactive tele-science from remotely located experimental centers.

The education and training of geoscientists in high performance computing and communication is being addressed through a broad spectrum of courses offered at NCAR and the other NSF supercomputer centers and through graduate assistantships and post-doctoral positions.

Technologies are being developed for Internet distribution of real-time geophysical data which has multiple sources at disparate locations. It is the aim of this effort to allow a university user transparent access to petabytes of geosciences data located at several centers from which data of interest could be selected.

Agency Ties	
ARPA	
NSF	
DOE	
NASA	Partner
NIH	
NSA	
NIST	
NOAA	Partner
EPA	
ED	

FY 1993 Accomplishments	FY 1994 Major Milestones	FY 1995 Major Milestones	Milestone Changes from Prior IP
Acquisition, maintenance, and support of SP-1 and CM-5 MPP systems. Develop, operate and maintain a clustered array of high performance workstations coupled together with high speed networking and a HIPPI switch technology. Support to users for use of MPPs and software conversion, including conversion of chemical cycle and cloud physics models and improvement of parallel algorithms for climate models. Support for two Postdoctoral Fellows.	Major upgrade to MPP computing by replacing the SP-1 with a 32 node SP-2. Support continued geophysical model conversion to MPP computers and continued development of parallel algorithms and techniques for effective use of MPP machines on geophysical codes. Demonstrate the 256 Kbps data transmission for between the US and South Pole Station. Beta test internet data distribution of real time data environmental data.	Enhance MPP computers available at NCAR by the installation of 256 node T3D. Use the Y-MP/T3D computers exclusively for climate simulation. Demonstrate Earth Simulation models simultaneously running on MPP, clustered and shared memory computers and connected by flux coupler software and high speed networking technology. Operation of 1.5 Mb/s link at Palmer Station, Antarctica using ACTS satellite. Partial implementation of internet data distribution.	Shift away from CM-5 to SP and T3D architectures on the basis of extensive performance testing with geophysical codes. The development and use of internet for data transmission to replace satellite delivery.

NSF	Budget Account Code	Budget (BA, $ M)				HPCC Components by 1995 Pres. Req.				
		FY 93 Actual	FY 94 Pres.	FY 94 Est.	FY 95 Pres.	HPCS	NREN	ASTA	IITA	BRHR
Computational Mathematics (non-NC/GC)		5.72	6.22	6.68	7.59			1.89	0.70	5.00

Mathematics plays a central role in the drive to produce faster and more accurate algorithms which, in tandem with hardware advances, produce state-of-the-art simulations across the wide spectrum of the sciences. Research supported under this activity satisfies at least one of the following criteria:

Use of high-performance computer systems and architectures as a testbed for research.

Innovative approaches to development of new algorithms, especially involving parallel and distributed environments.

Information-intensive activities, such as wide-area data exchanges, interaction services, and electronic collaborations.

Unusual developments involving symbolic and numeric computation.

Mathematical questions involving the preparation of suitable tools for visualization.

Agency Ties	
ARPA	
NSF	
DOE	
NASA	
NIH	
NSA	
NIST	
NOAA	
EPA	
ED	

FY 1993 Accomplishments	FY 1994 Major Milestones	FY 1995 Major Milestones	Milestone Changes from Prior IP
Developed novel expansions of visualization techniques and software tools for geometers and other mathematics researchers at the Geometry Center. Electronic information dissemination and environments for the mathematical sciences community were expanded. A program of theory, testing and use of random number generators for parallel processing was initiated. The linear algebra library LAPACK was extended to take advantage of high-performance computers.	Software developed under this program such as LAPACK will be distributed, upgraded and maintained. Connectivity and modes of accessing and sharing data and resources among the mathematical sciences community will be expanded. Building on user-friendly interfaces, workable and usable codes for group theorists, biologists, and other scientists will be developed targeted at high performance computing systems.	Software with a user-friendly interface will be developed and distributed to provide information on and to solve problems involving finitely-presented groups. A research group in mathematical biology will produce and distribute algorithms and software to solve problems in molecular sequence analysis, especially for analysis of DNA, RNA, and protein sequences. Algorithms and supported software for problems in fluid flow will be developed with particular emphasis on improved versions of Peskin's "immersed boundary method" for incompressible fluid dynamics in regions with complicated moving boundaries. Examples of applicable problems are blood flow in a beating heart and the structure of the earth in seismic oil exploration.	None.

NSF	Budget Account Code	Budget (BA, $ M)				HPCC Components by 1995 Pres. Req.				
Physical Sciences (non-NC/GC)		FY 93 Actual	FY 94 Pres.	FY 94 Est.	FY 95 Pres.	HPCS	NREN	ASTA	IITA	BRHR
		2.75	6.35	5.32	9.77			3.77	4.00	2.00

This activity provides support for research on fundamental problems in Astronomy, Chemistry, Material Science, and Physics using state of the art vector supercomputers or emerging massively parallel systems. The emphasis is on developing models and solution techniques that provide for qualitative and quantitative improvements in the simulations. From these more accurate simulations the researchers are able to gain new insights into the nature of the physical phenomena being simulated. In addition, this activity broadens the base of users of advanced computing systems by exposing graduate students and post doctoral researchers to the benefits of computational science as an intrinsic part of the scientific method.

Agency Ties	
ARPA	
NSF	
DOE	
NASA	
NIH	
NSA	
NIST	
NOAA	
EPA	
ED	

FY 1993 Accomplishments	FY 1994 Major Milestones	FY 1995 Major Milestones	Milestone Changes from Prior IP
Parallelized Cell Multipole Molecular Dynamics code on the KSR and demonstration of a practical million atom simulation for the surface tension of Teflon. Parallelized Pseudospectral Generalized Valence Bond (PS-GVB) technique for performing ab initio molecular orbital electronic structure calculations. Carried out computational studies involving the formation of elementary particles, the structure of the nucleus, the characterization of complex and chaotic phenomena, correlations in atoms and molecules, the equations of general relativity and the movement and analysis of data in physics experiments using MPPs and distributed computing paradigms. Implemented new diagonalization routine in a quantum mole. dynam. code which is being used to simulate small silicon clusters and liquids.	Parallelize the PRISM algorithm for integral calculations in GAUSSIAN suite of ab initio molecular orbital programs for the Cray T3D. Optimize the parallelized version of the PS-GVB software for the CM5 and Intel Paragon computers, and port it to the IBM SP2 and Cray T3D. Real-time acquisition, analysis and distribution of the data from particle accelerator experiments will be demonstrated using new software and networking capabilities. Homopolar semiconductor liquids such as germanium and heteropolar liquids such as gallium arsenide will be simulated using the improved quantum molecular dynamics code. Also, simple metals (sodium, potassium) will be modeled along with intrinsic defects.	Demonstrate the efficacy of parallelized CMM/MD codes by using them to predict glass transition temperatures of polymer blends and copolymers. Extension of the Newton-Euler Inverse Mass Operator method to fast dynamics on periodic systems with up to one million atoms. A new project to enable scientists in physics and other disciplines to acquire and distribute information easily and rapidly in the form of electronic preprints will be initiated. The CARM program that supports high performance computing research in the material sciences will be evaluated and expanded or refocused.	None.

NSF	Budget Account Code	Budget (BA, $ M)				HPCC Components by 1995 Pres. Req.				
		FY 93 Actual	FY 94 Pres.	FY 94 Est.	FY 95 Pres.	HPCS	NREN	ASTA	IITA	BRHR
Social, Behavioral & Economic Sciences (non-NC/GC)		0.65	1.00	1.15	3.01			0.67	2.34	

The basic research division of SBE (The Division of Social, Behavioral and Economic Research) was created at the beginning of 1993, by amalgamation of many existing programs. SBER is currently engaged in the development of a general plan for computing in the social and behavioral sciences, and it is examining how best to contribute to the new HPCC component, IITA. There are five program clusters in SBER, and each one has held a workshop on aspects of high performance computing that relate to its disciplines:

1. Cognitive, Psychological, and Language Sciences cluster: "Cognitive Science" workshop.

2. Anthropological and Geographic Sciences cluster: "Computational Geography" workshop.

3. Economics, Decision and Management Sciences cluster: "Computational Economics" workshop.

4. Social and Political Sciences cluster: Artificial Social Intelligence" workshop.

5. Science, Technology and Society cluster: "Electronic Networks" workshop.

SBE will be heavily involved in the new IITA component, because its goals are close to the social and behavioral sciences.

Agency Ties	
ARPA	
NSF	
DOE	
NASA	
NIH	
NSA	
NIST	
NOAA	
EPA	
ED	

FY 1993 Accomplishments	FY 1994 Major Milestones	FY 1995 Major Milestones	Milestone Changes from Prior IP
Demonstration projects establish feasibility of combining artificial intelligence simulations with experimental laboratory studies of human social and economic interaction. Progress in geographic information systems (GIS) through supporting the National Center for Geographic Information and Analysis. Completed review of needs and opportunities for advanced computing in the social, behavioral and economic sciences.	Conduct comprehensive review of software tools required to manage large socio-economic databases as part of the National Information Infrastructure. Establish systematic plan to develop methods of advanced computing in the social, behavioral and economic sciences. Continue exploration of scientific opportunities for artificial intelligence in psychology, sociology and economics.	Establish distributed processing geographic information systems for the National Information Infrastructure. Adapt of genetic algorithms, neural networks and symbolic processors to a wide range of problems in the social sciences. Demonstrate new methods for providing wide access to previously unavailable government statistics, while preserving data integrity and preventing misuse.	None.

NSF	Budget Account Code	Budget (BA, $ M)				HPCC Components by 1995 Pres. Req.				
Education and Training		FY 93 Actual	FY 94 Pres.	FY 94 Est.	FY 95 Pres.	HPCS	NREN	ASTA	IITA	BRHR
		15.34	19.90	20.04	20.24		9.00			11.24

The education and training effort in the HPCC consists of existing activities that focus more directly on HPCC and of new activities begun as part of the HPCC initiative. Many of these activities are interdisciplinary in nature drawing on the resources of all of NSF's directorates. In particular, the pilot testbeds projects are collaborative projects between the EHR directorate and the CISE directorate drawing on scientific expertise from throughout NSF. Specific examples follow:

MOSIS - An existing activity training students and providing research infrastructure for the design and manufacture of custom VLSI chips.

Undergraduate Education - An existing activity which provides some funding for new HPCC course and curriculum development.

Research Experiences for Undergraduates - Expanded to provide more opportunities for undergraduates to perform research in HPCC areas.

Superquest - An expanded activity that provides for high school student research experiences at Supercomputer Centers.

Postdoctoral Research Associates - A new activity to provide postdoctoral training in computational science and engineering and experimental computer science.

Pilot Educational Networks - Develop networks to develop, implement, test, and evaluate applications of computer and communications to education. \

Network Infrastructure for Education - A new joint EHR/CISE activity, starting in FY 1994, that will explore issues of large-scale networking uses in education.

Agency Ties	
ARPA	Partner
NSF	
DOE	
NASA	
NIH	
NSA	
NIST	
NOAA	
EPA	
ED	

FY 1993 Accomplishments	FY 1994 Major Milestones	FY 1995 Major Milestones	Milestone Changes from Prior IP
K-12 Accomplishments: Super- quest funded 5 sites with 4 teams of 4 high school students and 1 teacher at each site to perform research in high performance computing. Pilot education networks funded four coalitions which developed pilot networks ranging from citywide to nationwide for K-12 education. Undergraduate Education Accomplishments: MOSIS - a service provided to universities for research and training in VLSI design and fabrication - was used in 175 universities in 47 states by approximately 5,000 students. Research Experiences for Undergraduates is an ongoing NSF activity that has seen significant participation by the supercomputer centers and the science and technology centers. In 1993 approximately 360 students were supported by this program.	Enhancement of existing networks to demonstrate scalability and effectiveness of applications of networks to the infrastructure of education.	Development of new testbeds to demonstrate scalability, cost, policy, effective applications and other characteristics of the successful use of computer networks in education.	New infrastructure activities in networked applications in education.

NSF	Budget Account Code	Budget (BA, $ M)				HPCC Components by 1995 Pres. Req.				
		FY 93 Actual	FY 94 Pres.	FY 94 Est.	FY 95 Pres.	HPCS	NREN	ASTA	IITA	BRHR
IITA Program			36.00	11.42	35.25				35.25	

NSF's IITA program for FY95 totals $50.64M; $15.39M of the total is distributed to other Program Activity Components and is thus not included in the total for this component. However, the description of this component is intended to cover the entire IITA activity. NSF's IITA research program develops new technologies and applications that make HPCC more broadly applicable. While the other HPCC components focus on advances in science and engineering, IITA extends the scope of the HPCC by making the high performance resources and infrastructure available to a dramatically larger community of users. To achieve these goals, IITA will support work leading to (1) the development of ubiquitous information services, (2) the development of new software and support environment for broad access to distributed resources, (3) the development of intelligent interfaces, and (4) the demonstration of HPCC's applicability in solving National Challenges in such areas ad Digital Libraries, Civil Infrastructure, Manufacturing, Education & Lifelong Learning, and Government Information Delivery. Ongoing IITA activities include:

- Information Infra. Services - New program emphases will be placed on the enhancement of communication and computing services tailored to the general public.
- Intelligent Interfaces - The goal is to create and insert an intelligent service layer that will significantly broaden the base of information providers, developers, and consumers, while reducing the existing barriers to accessing and using information for real world applications.
- System Development and Support Environment - The goal is to create a more effective software development paradigm and technology base founded on the principles of composition rather than construction, solid architectures rather than ad hoc styles, and more direct user involvement in all stages of the software life cycles.
- National Challenges - The entire technology base developed in IITA will be leveraged across the National Challenges, leading to significant economies of scale in the development costs. NSF leads a new cross agency (with ARPA, NASA) research initiative in Digital Libraries, beginning in FY 1994. The research will focus on a wide range of areas designed to lay the foundation for and demonstrate the next generation of knowledge products and Information services.

NSF also participates in several other new initiatives under the IITA program: Civil Infrastructure, Advanced Manufacturing, Networking Infrastructure for Education, and Government Information Delivery.

Agency Ties	
ARPA	Partner
NSF	
DOE	
NASA	Partner
NIH	
NSA	
NIST	
NOAA	
EPA	
ED	

FY 1993 Accomplishments	FY 1994 Major Milestones	FY 1995 Major Milestones	Milestone Changes from Prior IP
Development and dissemination of Mosaic, a client/server software which discovers, retrieves, and displays multimedia documents over the Internet, by NSF's Supercomputer Center at the University of Illinois. Mosaic has been widely used by government agencies for delivery of information and currently has over one million public users. Implementation of a research initiative on Human Language Technology, jointly with ARPA. Development of a cross agency research initiative on Digital Libraries (with ARPA and NASA). Sponsorship of a series of workshops and studies to develop research agenda for National Challenge Applications in manufacturing, virtual environments, and computer assisted surgery (health care).	Establish six major research teams to pursue work in Digital Libraries each to develop an experimental testbed. Develop a new research thrust in Virtual Prototyping for Mfg. to accelerate progress in applying information technology and distributed computing and communications capabilities. Develop a new research thrust in Problem Solving Environments to stimulate new work in software development that's tailored to problem domains, user needs, and emphasis on composition and reuse. Participate in the new program on Networking Infra. for Education to accelerate the adoption of adv. information technologies for science and mathematics education. Demonstrate a multi-media pilot system for delivering government information to the public.	Develop a supplemental program for 25 additional institutions to participate in the Digital Libraries initiative. Demonstrate preliminary models and data developed by the research teams under the cross agency Digital Libraries initiative. Implement a cross-disciplinary research program on Virtual prototyping for Advanced Mfg. Develop and implement a program on Reliable Computer Controlled Systems to accelerate progress in providing safe and reliable computing and commun- ication devices and systems for National Challenge application areas. Develop laboratories as part of the new institutional infrastructure needed for investigating effective human utilization of massive information stores.	None.

Department of Energy

The Department of Energy (DOE) HPCC program is a diverse applied mathematical sciences and high performance computing and communications program which spans the spectrum of activities from strategic fundamental research to technology development. Its mission is:

> To improve the ability of DOE to solve scientific and engineering problems which are critical to its mission, to the NII Agenda for Action, to the Energy Policy Act (EPACT), and to the national interest, through research and development in and applications of advanced mathematical, computational, computer, and communications sciences and information technologies and, as authorized, to meet the requirements of the HPC Act of 1991 and the Administration's program for the NII: Agenda for Action.

Because of the computationally intensive nature of energy related applications and problems, DOE depends on advancements in computational techniques and computer and networking technologies to accomplish its mission. As a result, DOE has a long history of computational research and development, with strong industrial and university cooperation. DOE's Applied Mathematical Sciences (AMS) Program (a forerunner of the HPCC program) was initiated in the early nineteen fifties to enhance understanding of the use of digital computers in nuclear sciences and other energy research applications. DOE has been prominent in maintaining the U.S. leadership in high performance computing and communications, in encouraging and providing innovation in HPCC and other information technologies, and in supporting U.S. economic, scientific, and technological competitiveness and productivity through its extensive use of these technologies.

The DOE HPCC program covers all five components of the HPCCI. Its original HPCC Program results from DOE's role as a pioneer in the use of supercomputers for numerical simulation of complex systems. This program reduces substantially the cost of experimentation and testing and also reduces the adverse effects on the environment. The revised HPCC Program includes R&D in advanced communications technologies which could radically transform the way in which Americans work. The most important economic benefits of the HPCC Program may well result from its newest initiative in IITA. The R&D activities in high performance computing technologies, including telecommuting (or telework in general), are also one of the EPACT objectives. These R&D efforts are aimed at advancing the enabling technologies for telecommuting and other telework which will provide for reduced oil dependency, reduced pollutants, better productivity for disabled and other workers, and other societal benefits. In the nearer term these activities are critical to DOE's efforts to provide geographically distributed access to DOE's many important experimental facilities and computational resources.

The Office of Scientific Computing (OSC) manages DOE's HPCC program and serves as the advocate and lead within the Department to formulate and coordinate the Department's National Information Infrastructure (NII) initiative. The HPCC Program supports the underlying mathematical concepts and information technology needs of all DOE mission areas (e.g., Defense, Energy Efficiency, Environmental and Fossil programs, etc.). OSC's plan promotes U.S. economic and technological competitiveness in many critical technology areas and involves many industrial partnerships. The DOE collaborates closely with other agencies which are HPCCI users, such as NASA, to produce software which will be generally useful. The DOE HPCC program also complements ARPA basic technology research and NSF basic research programs. Within DOE, OSC manages and provides communications and computational resources for investigators supported by Energy Research. Users meetings (e.g., Energy Research Supercomputer Users Group, ESnet Steering Committee and ESnet Site Coordinator Committee) help ensure that the developing technologies will support the DOE mission. In jointly funded projects (with other Energy Research Offices) such as the Grand Challenge applications projects, OSC maintains active communications with partners for selecting projects and monitoring progress. In addition, OSC provides primary support to DOE's representative on the Administration's Information Infrastructure Task Force.

Department of Energy

Program Activity	Budget Account Code	Partner/User Agencies	Budget (BA, $ M)				HPCC Components by 1995 Pres. Req.				
			FY 93 Actual	FY 94 Pres.	FY 94 Est.	FY 95 Pres.	HPCS	NREN	ASTA	IITA	BRHR
Advanced Prototype Systems	KC 07 01	ARPA, NASA		1.00	1.00	1.00	1.00				
Evaluation of Early Systems	KC 07 01, KC 07 02, KC 35 07	ARPA	7.73	9.90	9.80	9.90	9.90				
Energy Sciences Network (ESnet)	KC 07 02 KC 35 07	ARPA, NSF, NASA, ...	7.68	14.80	14.60	14.80		14.80			
Gigabit Research and Development	KC 07 02	ARPA, NSF, NASA	1.86	2.00	2.00	2.00		2.00			
Enabling Energy Grand Challenges	KC 07 01	NSF, NASA	6.53	9.00	8.90	9.00			9.00		
Software Components and Tools	KC 07 01	NASA	8.98	12.60	12.50	12.60			12.60		
Computational Techniques	KC 07 01	EPA	2.44	3.40	3.30	3.40			3.40		
Supercomputer Access	KC 07 02, KC 35 07		35.13	35.40	34.80	35.60			35.60		
HPC Research Centers	KC 07 01, KC 07 02, KC 35 07		7.15	12.50	12.30	12.90			12.90		
HPC Research Centers-Global Climate Collaboration	KP 05		1.80	2.00	2.00	2.00			2.00		
Information Infraservices	KC 07 02			0.30	0.30	1.20				1.20	
Research Participation and Training	KC 07 01		2.60	2.00	2.00	2.00					2.00
Education, Training and Curriculum	KC 07 01		2.36	3.00	3.00	3.00					3.00
Basic Research for Applied Mathematics Research	KC 07 01	NSF	15.25	15.70	15.70	16.00					16.00
		Totals:	99.51	123.60	122.20	125.40	10.90	16.80	75.50	1.20	21.00

DOE	Budget Account Code	Budget (BA, $ M)				HPCC Components by 1995 Pres. Req.				
		FY 93 Actual	FY 94 Pres.	FY 94 Est.	FY 95 Pres.	HPCS	NREN	ASTA	IITA	BRHR
Advanced Prototype Systems			1.00	1.00	1.00	1.00				

This activity funds research into advanced computer architectures and novel commercial prototypes suitable for scientific computing. Funding for development of new prototype architectures has been deemphasized in favor of providing feedback to commercial vendors of novel systems. This activity does include funding to develop a message passing interface for an MPP system, which is critical for easily porting codes among massively parallel computers.

Agency Ties	
ARPA	Partner
NSF	
DOE	
NASA	Partner
NIH	
NSA	
NIST	
NOAA	
EPA	
ED	

FY 1993 Accomplishments	FY 1994 Major Milestones	FY 1995 Major Milestones	Milestone Changes from Prior IP
The KSR-1 machine located at the Partnership In Computational Science (PICS) HPCRC at Oak Ridge National Laboratory was acquired jointly by DOE and ARPA.	Lease IBM SP-1 (in cooperation with ARPA) at Argonne National Laboratory (ANL). Initiate joint agency project in enhanced I/O for Parallel computers.	Produce the reference implemen- tation for a message-passing system interface on massively parallel processors (deliverable from IBM/ANL collaboration using SP-1). Solve the Kajiya rendering equation for scalable parallel systems at extremely high speed and fidelity to put a firm scientific basis under the Art of performance analysis for parallel systems.	

DOE	Budget Account Code	Budget (BA, $ M)				HPCC Components by 1995 Pres. Req.				
		FY 93 Actual	FY 94 Pres.	FY 94 Est.	FY 95 Pres.	HPCS	NREN	ASTA	IITA	BRHR
Evaluation of Early Systems		7.73	9.90	9.80	9.90	9.90				

This program activity helps fund the acquisition of large prototype computers for testing and evaluation. Grand Challenge-class problems require large prototypes because they cannot be scaled down without removing essential aspects of their physics. This activity covers:

* identifying candidate systems
* acquiring large prototype systems (with co-sponsors, where possible)
* placing the systems at sites with access to Internet

Agency Ties	
ARPA	Partner
NSF	
DOE	
NASA	
NIH	
NSA	
NIST	
NOAA	
EPA	
ED	

FY 1993 Accomplishments	FY 1994 Major Milestones	FY 1995 Major Milestones	Milestone Changes from Prior IP
Machines acquired (at least partially) under this program activity include: an Intel Touchstone Delta iPSC/860 (Caltech), Thinking Machines CM5 at Los Alamos National Laboratory (LANL), a Kendall Square KSR-1 (in cooperation with ARPA) at Oak Ridge National Laboratory (ORNL), and a 512 processor Intel Paragon Prototype Multiprocessor at ORNL. No new advanced prototype systems are likely to be acquired in FY 1995.	Upgrade the Intel Paragon to 1024 processors. Convene an Industrial Computing Requirements Forum led by Ames Laboratory. Bring together computer technology vendors, code developers, and applications scientists to reduce design cycles for computing equipment and applications software.	Upgrade the Intel Paragon to 2000 processors. Have full-scale CM-5 and Paragon systems operational and supporting Grand Challenge research projects.	

DOE	Budget Account Code	Budget (BA, $ M)				HPCC Components by 1995 Pres. Req.				
		FY 93 Actual	FY 94 Pres.	FY 94 Est.	FY 95 Pres.	HPCS	NREN	ASTA	IITA	BRHR
Energy Sciences Network (ESnet)		7.68	14.80	14.60	14.80		14.80			

ESnet is DOE's network component of the National Research and Education Network (NREN) and Internet. It provides worldwide connectivity for the scientific community to Energy Research facilities, including: Light Sources, Neutron Sources, Particle Accelerators, Spectrometers, Particle Accelerators, High Performance Computing Research Centers (HPCRC), Genome Centers, data banks, and other leading-edge science instruments and facilities. ESnet provides advanced services through the early acquisition and testing of commercially supplied communications and network services. Funding covers acquisition of services from communications companies (basic bandwidth for normal communications, primary rate ISDN for video, video multiports/ services), hardware to expand network capabilities (routers, international links), personnel to develop tools for network use (e.g., communication, authentication, and privacy protocols), and maintenance.

Agency Ties	
ARPA	Partner
NSF	Partner
DOE	
NASA	Partner
NIH	
NSA	
NIST	User
NOAA	
EPA	User
ED	

FY 1993 Accomplishments	FY 1994 Major Milestones	FY 1995 Major Milestones	Milestone Changes from Prior IP
Implemented prototype x.500 directory service. Provided international connections to support Energy Research Programs.	Begin upgrade of ESnet to 45 Mbs fast packet services. Extend video/voice conferencing capabilities to permit transmission of conventional video systems over ESnet and dedicate a portion of ESnet for packetized video conferencing. Investigate connection of wide area ESnet Asynchronous Transfer Mode (ATM) services to each site's local area networks. Continue the KERBEROS-based authentication pilot project to investigate technical and administrative issues for deploying national scale authentication. Begin to deploy new KERBEROS authentication services to five sites out of roughly 25 total.	Complete ESnet upgrade to 45 Mbs fast packet services. Initiate the upgrade of a few sites to 155 Mbs to three or four sites. Upgrade ESnet links to Germany (T1), Italy (T1), and Russia (128 Kbs) to support multi-national science collaborations, such as the International Thermonuclear Engineering Reactor. Provide key types of video and collab. work space capabilities within ESnet, including production quality video/voice capabilities. The choices will accommodate different cost/performance options and interoperability needs. Complete the KERBEROS Authentication pilot project and make these services available to the ESnet community.	Upgrade of ESnet to 45 Mbs has slipped two years (from 93 to 95) due to procurement delays and funding constraints. Initial upgrades to 155 Mbs have been deferred from 94 to 95. More detailed milestones added since previous IP.

DOE	Budget Account Code	Budget (BA, $ M)				HPCC Components by 1995 Pres. Req.				
		FY 93 Actual	FY 94 Pres.	FY 94 Est.	FY 95 Pres.	HPCS	NREN	ASTA	IITA	BRHR
Gigabit Research and Development		1.86	2.00	2.00	2.00		2.00			

This component funds the gigabit and advanced networking research activities of the NREN. It includes gigabit test bed activities (HIPPI, CASA, BLANCA, MAGIC, BAGABITS, National Storage Lab, Electronic Telescope, and more), packetized video and voice, collaborative workspace, telecommuting projects, and scalable communications libraries for distributed memory computing environments that offers the user a Single Program Multiple Data programming paradigm.

Funding will be provided for research and development for ATM issues with regard to High Performance Parallel Interface (HIPPI), standards implementation, and interfaces. Funding will also be used to investigate the network resource reservation and management for isochronous traffic. Further work in security, such as incorporating KERBEROS into Wide Area Information Services (WAIS) for secure info retrieval/search and interoperable authentications realms, intelligent user interfaces for accessing the network, interoperable video systems, high speed LANs, telecommuting, and collaborative work environments will be funded. In addition, work needs to begin on identifying the real issues with inter-network peering and routing. These issues will not be completely nor properly addressed in the NSF routing arbiter solicitation and they are crucial to the continued operation of the Internet. Research on routing and addressing (e.g., end system identifiers, network versus geographical addressing, automatic host configuration, etc.) for the growth of the Internet on a multi-protocol dimension needs to be conducted.

Agency Ties	
ARPA	Partner
NSF	Partner
DOE	
NASA	Partner
NIH	
NSA	
NIST	
NOAA	
EPA	
ED	

FY 1993 Accomplishments	FY 1994 Major Milestones	FY 1995 Major Milestones	Milestone Changes from Prior IP
The packetized video/voice and collaborative workspace (whiteboard) are being used by many on the Internet and forms a major basis of the available tools for interactive voice/video. The Internal Engineering Task Force video-cost conferences and other Internet interactive activities require the tools developed at Lawrence Berkeley Laboratory (LBL) and other DOE sites.	Video Audio Transport (VAT) and Whiteboard are being used on the network now. MBONE, the virtual Multicast backbone used to support IP based audio visual traffic, is a major success, with the IETF meetings being regularly videocast. Video, output from commercial codecs and encapsulated in IP packets, will be used across ESnet. This work is being done in cooperation with the commercial codec vendor who is expected to market the results in the near future. HIPPI is the dominant HPCS interface to networks. Two DOE telecommuting studies will be released this year.	Scalable communications Library will be completed by 6/95 LBL will be releasing beta code of collaborative technologies to the Advanced Research Projects Agency (ARPA) and other agencies for further refinement and development Packetization of commercial codec video/audio output will be completed and will allow ESnet to carry packetized video encapsulated in IP	

DOE	Budget Account Code	Budget (BA, $ M)				HPCC Components by 1995 Pres. Req.				
		FY 93 Actual	FY 94 Pres.	FY 94 Est.	FY 95 Pres.	HPCS	NREN	ASTA	IITA	BRHR
Enabling Energy Grand Challenges		6.53	9.00	8.90	9.00			9.00		

In FY 1992, DOE initiated nine projects that support Grand Challenges critical to energy planning. All projects are cofunded by other DOE programs and by industrial partners. Participants include DOE laboratories, universities, industry and other HPCC agencies. These projects construct prototype solutions through the development of advanced algorithms and software and the use of HPCC resources. Selection of these projects was made by a panel including DOE program managers and other HPCC agency participants. Each project undergoes periodic reviews to assess research progress and future plans for continued funding. The projects are:

1. Computational Chemistry (CC) - parallelize key chemistry codes that permit researchers to study environment problems.
2. Computational Structural Biology - develop methods for modeling components of genomes and a parallel programming environment for structural biology.
3. Mathematical Combustion Modeling (MCM) - develop adaptive parallel algorithms for computational fluid dynamics and apply these methods to key private sector and government combustion models.
4. Quantum Chromodynamics Calculations - develop lattice gauge algorithms on massively parallel machines for high energy and particle physics applications.
5. Oil Reservoir Modeling - construct efficient algs. for parallel systems to model fluid flow through permeable media for better oil recovery methods from wells.
6. The Numerical Tokamak Project - develop and integrate particle and fluid plasma models on MPCs as part of a study of Tokamak fusion reactors.
7. Global Climate Modeling (GCM) - develop and implement versions of large-scale atmosphere and ocean general circulation models for MPCs.
8. Groundwater Transport and Remediation (GTR) - design and implement a multiphase groundwater transport code with interface tracking, fracture flow, geochemistry, biodegradation and colloidal microtransport.
9. First Principles Simulation of Materials Properties - develop scalable parallel algorithms for performing local density approximation simulations of materials to guide the discovery of advanced materials with novel properties.

Agency Ties	
ARPA	
NSF	Partner
DOE	
NASA	Partner
NIH	
NSA	
NIST	
NOAA	
EPA	
ED	

FY 1993 Accomplishments	FY 1994 Major Milestones	FY 1995 Major Milestones	Milestone Changes from Prior IP
Projects initiated	CC - implement parallel code for self-consistent field (SCF) and second order many-body perturbation theory (MP2) electronic structure calculations that incorporate fully distributed data structures. MCM - make available a production quality code capable of simulating low-speed 3D combustion flows in simple geometries used in designing commercial burners. GCM - optimize the Sempter- Chervin global ocean model to perform coupled ocean-atmosphere simulations that allow, among other things, the prediction of large-scale eddies in the Gulf of Mexico. GTR - Add to the transport code capabilities for 2D front tracking, subsurface heterogeneity estimation, and large- and small-scale reservoir flow resolution.	CC - develop scalable nonlinear optimization techniques and enhance trial functions and solution schemes for the SCF and MP2 codes. MCM - extend code to simulate low-speed, 3D combustion flows in complex geometries like those found in internal combustion engines. GCM - test an improved particle-in-cell method for the parallel global ocean model, apply differential sensitivity analysis, and employ a barotropic closure model in resolving large-scale ocean dynamics. GTR - test bioremediation and radionuclide decay strategies and benchmark two-phase flow simulations, including results from parameter estimations.	

DOE

Software Components and Tools

Budget Account Code	FY 93 Actual	FY 94 Pres.	FY 94 Est.	FY 95 Pres.	HPCS	NREN	ASTA	IITA	BRHR
	8.98	12.60	12.50	12.60			12.60		

Budget (BA, $ M); HPCC Components by 1995 Pres. Req.

Developing advanced software to facilitate the use of high performance systems to scientific problems, including:

* Constructing, maintaining, and monitoring scientific software
* Understanding and managing the input and output of scientific software
* Facilitating and enhancing scientific and engineering collaborations.

This program is focused on fulfilling the needs of users who are a critical part of the review process.

Agency Ties	
ARPA	
NSF	
DOE	
NASA	Partner
NIH	
NSA	
NIST	
NOAA	
EPA	
ED	

FY 1993 Accomplishments	FY 1994 Major Milestones	FY 1995 Major Milestones	Milestone Changes from Prior IP
Performance monitoring tools in PCN at Argonne National Laboratory; work on parallel software monitoring tools at LANL, and ORNL and University of Wisconsin; and work in object oriented programming at Sandia National Laboratory, Lawrence Berkeley Laboratory, and Purdue. Work at Rutgers on automatic identification of and tracking of features in large data sets such as islands in MHD simulations; a collaborative effort between LBL, LLNL, and University of Maryland to understand how to store and retrieve massive data sets with spatial structure, e.g., resulting from global change research, in hierarchical storage systems; and work in advanced visualization techniques and algorithms at a number of DOE labs.	Evaluate near and long-term strategies for developing portable codes and benchmarks for parallel computers using user-scale codes Evaluate the scalability of parallel operating systems, I/O and message passing schemes for real user applications	Establish distributed computing and experimental collaboration testbed for evaluating security, resource management, program monitoring, and other computational science tools in this environment. Develop flexible, interoperable tools for portable parallel program creation, testing and performance monitoring, and interpretation of output on advanced architectures Investigate ways to involve U.S. industry with Grand Challenge research teams to improve economic competitiveness	

DOE	Budget Account Code	Budget (BA, $ M)				HPCC Components by 1995 Pres. Req.				
		FY 93 Actual	FY 94 Pres.	FY 94 Est.	FY 95 Pres.	HPCS	NREN	ASTA	IITA	BRHR
Computational Techniques		2.44	3.40	3.30	3.40			3.40		

Continue application of innovative computational mathematics and computer science to key scientific and engineering applications, including: computational design of materials and molecules; global ocean modeling; and algorithm and grid design for both solid and fluid mechanics problems. Emphasize resolving algorithmic bottlenecks to effectively using high performance systems to solve full-scale application problems. This activity is focused on delivering robust versions of advanced algorithms that are important for Grand Challenge Problems: improved Fast Forrier Transforms, molecular dynamics solvers, particle pushers, etc.

Agency Ties	
ARPA	
NSF	
DOE	
NASA	
NIH	
NSA	
NIST	
NOAA	
EPA	Partner
ED	

FY 1993 Accomplishments	FY 1994 Major Milestones	FY 1995 Major Milestones	Milestone Changes from Prior IP
The computational techniques activity has fed the development of improved simulations in global ocean modeling and in the design of materials and molecules. It has also served as an incubator for the creation of new Grand Challenges. Optimized parallel FFT software releases.	External evaluation of the activity on computational design of materials and molecules Initiate research on common problems in computational aerodynamics and global ocean modeling Test advanced mesh generation techniques in support of both solid and fluid dynamics applications	Develop techniques for algorithmic analysis of grand challenge problems	

DOE	Budget Account Code	Budget (BA, $ M)				HPCC Components by 1995 Pres. Req.				
Supercomputer Access		FY 93 Actual	FY 94 Pres.	FY 94 Est.	FY 95 Pres.	HPCS	NREN	ASTA	IITA	BRHR
		35.13	35.40	34.80	35.60			35.60		

Funding for equipment and personnel for Supercomputer Access Program at the National Energy Research Supercomputer Center (NERSC). Provide high performance production computing for investigators supported by the Office of Energy Research. Mission areas include: Materials, Chemistry, Geosciences, Biosciences, Engineering, Health and Environmental Research, High Energy and Nuclear Physics, Fusion Energy, applied mathematics and computational science. The Center serves more than 4000 users working on about 700 projects; about 35% University-based, 60% in National Laboratories and 5% in industry. It operates 4 CRAY computers: a C-90 with 16 processors, with 256 Million words (Mw) of memory, a CRAY-2 with 8 processors with 128 Mw, and a CRAY-2 with 4 processors and 128 Mw, and the National Education Supercomputer, a single processor Cray X-MP donated by Cray Research, which is available over the Internet to high schools for educational programs.

Agency Ties	
ARPA	
NSF	
DOE	
NASA	
NIH	
NSA	
NIST	
NOAA	
EPA	
ED	

FY 1993 Accomplishments	FY 1994 Major Milestones	FY 1995 Major Milestones	Milestone Changes from Prior IP
House the National Storage Laboratory and the follow-on, the High Performance Storage Systems, projects pertinent to one of the most severe limitations to high performance computing. The NSL was established by a CRADA among six industrial firms and NERSC. This collaboration currently involves several DOE laboratories, about a dozen U. S. storage vendors, and a university. The NSL will help serve the data storage needs of researchers at the DOE and at other Federal agencies and will help participating vendors develop new storage products. NERSC is also the site of a premier supercomputing educational program for high school students, the National High School Supercomputer Honors Program in Supercomputing, known as the Superkids program, which was begun in 1986.	The Special Parallel Processing Program expanded from 20,000 Computer Resource Units (CRUs) to 60,000 by 4Q94. Central User Bank (CUB) brought into production in 1Q94. This provides a single accounting for usage of the three production Crays at the center and the gives the user the ability to shift resources to the machine most optimal for her purposes. Institute storage quota system in 1Q94. It provides better utilization of the most scarce of resources at the Center, the Central File System and other storage. These improvements and other efficiencies provided a 30% increase in CRUs delivered to ER programs. The NSL will produce a beta version of NSL Unitree by 4Q94. Initiate procurement of MPP system for production use.	Implement massively parallel computer in early production environment.	

DOE	Budget Account Code	Budget (BA, $ M)				HPCC Components by 1995 Pres. Req.				
		FY 93 Actual	FY 94 Pres.	FY 94 Est.	FY 95 Pres.	HPCS	NREN	ASTA	IITA	BRHR
HPC Research Centers		7.15	12.50	12.30	12.90			12.90		

Support the operation of two HPCRCs (at Oak Ridge and Los Alamos), which provide critical resources for enabling grand challenge applications. To this end the HPCRC's acquire large, full-scale, high performance computers exploiting innovative architectures for use by computational scientists working on Grand Challenge applications. The centers integrate these early high-performance computing systems into prototype heterogeneous computing configurations and make them available through the NREN. This allows for a wide spectrum of experiments for scalability studies, as well as an opportunity to provide grand challenge researchers access to the largest possible advanced systems. To carry out this program the centers also support researchers in areas such as algorithm development, software components and tool development, mass storage, scientific visualization, basic research, and education programs.

Los Alamos National Laboratory operates a 1024-node CM-5 with 32 Gigabytes of memory and 128 Gigabytes of Scalable Disk Array using 4 HIPPI channels from Thinking Machines, Inc. This machine has a theoretical capacity of 128 Gigaflops and several codes have achieved 50 Gigaflops.

Oak Ridge National Laboratory operates an nCUBE 2 Model 640, a distributed memory hypercube machine used by high school students in the Adventures in Supercomputing program, a Kendall Square KSR1- 64 used to study shared memory algorithms, and an Intel Paragon XP/S 5, a 66-node 5 Gigaflops machine.

Agency Ties	
ARPA	
NSF	
DOE	
NASA	
NIH	
NSA	
NIST	
NOAA	
EPA	
ED	

FY 1993 Accomplishments	FY 1994 Major Milestones	FY 1995 Major Milestones	Milestone Changes from Prior IP
The Parallel Ocean Program, running with a grid of 1280 longitudinal points and 896 latitudinal points, has achieved enough resolution to be able to see Loop Current eddies passing into the Gulf of Mexico between Yucatan and Cuba. These currents can force the suspension of drilling operations on platforms in the Gulf. The Materials Science Grand Challenge was able to use the XP/S 5 to produce remarkable improvements in the resolution of the potential energy surface of a Germanium atom adsorbed onto Silicon.	Expand Parallel Ocean Program work to provide predictions to aid platform design and placement decisions.	150 Gigaflops machine at ORNL running in single user mode 4Q 1994 Upgrade machine at ORNL to multiple user/simple partition - 1Q 1995 Upgrade machine at ORNL to multiple user/overlap partition - 4Q 1995 Upgrade memory from 32 to 128 Gigabytes at LANL.	

DOE	Budget Account Code	Budget (BA, $ M)				HPCC Components by 1995 Pres. Req.				
		FY 93 Actual	FY 94 Pres.	FY 94 Est.	FY 95 Pres.	HPCS	NREN	ASTA	IITA	BRHR
HPC Research Centers-Global Climate Collaboration		1.80	2.00	2.00	2.00			2.00		

DOE's Office of Health and Environmental Research contributes direct support through its CHAMMP and Atmospheric Radiation Measurements programs and wastewater monitoring programs. These funds support grand challenge work in these areas at the two HPCRCs.

Agency Ties	
ARPA	
NSF	
DOE	
NASA	
NIH	
NSA	
NIST	
NOAA	
EPA	
ED	

FY 1993 Accomplishments	FY 1994 Major Milestones	FY 1995 Major Milestones	Milestone Changes from Prior IP
Initiate joint work in parallel ocean model and groundwater transport in collaboration with CHAMMP and other OHER programs.	Increase resolution of GCM to support global CO2 modeling. Direct experimental test of wastewater amelioration strategies developed in FY 1993	Continue R&D on Grand Challenge projects at the HPCRCs. These projects are cofunded by other DOE programs and by industrial partners. Participants include DOE laboratory, university, industry and other HPCC agency researchers. The are: * Global Climate Modeling, * Groundwater Transport and Remediation, and * First Principles Simulation of Materials Properties. Communicate results expected for some large applications.	

DOE	Budget Account Code	Budget (BA, $ M)				HPCC Components by 1995 Pres. Req.				
		FY 93 Actual	FY 94 Pres.	FY 94 Est.	FY 95 Pres.	HPCS	NREN	ASTA	IITA	BRHR
Information Infraservices			0.30	0.30	1.20				1.20	

This element supports research in the basic technology and services which underlie a number of IITA applications including energy demand management and environmental monitoring and waste minimization. The basic services being explored include authentication, distributed resource management, and distributed collaboration tools. Much of this work is done in close cooperation with U.S. computer manufacturers and potential users in industry.

Agency Ties	
ARPA	
NSF	
DOE	-
NASA	
NIH	
NSA	
NIST	
NOAA	
EPA	
ED	

FY 1993 Accomplishments	FY 1994 Major Milestones	FY 1995 Major Milestones	Milestone Changes from Prior IP
Covered under NREN	Establish security for remote access to resources (smart cards, authentication, etc.) such as libraries, digital and object databases, video conferences, robots in "hazardous areas," etc. Develop protocols and interfaces (GUI) for shared drawing/writing spaces, version/document control, a common spreadsheet interface, information, digital library, multimedia, object databases that must work at both high speeds and over a slow serial dialup line. Establish resource consumption monitoring tools and accounting tools.	Telework support (std file systems, graphics, spreadsheets, serial interfaces for access and configuration, sliding bar control of video interaction. Merge FAX/e-mail/phone/video (These tools/technologies need to work on many platforms and work at 2400-19.2, N-ISDN, and higher speed technologies to satisfy all levels of users.) Assess and document the benefits of and develop a plan for implementing an energy demand and supply management national challenge in close collaboration with the U.S. Power Utilities.	

DOE Research Participation and Training	Budget Account Code	Budget (BA, $ M)				HPCC Components by 1995 Pres. Req.				
		FY 93 Actual	FY 94 Pres.	FY 94 Est.	FY 95 Pres.	HPCS	NREN	ASTA	IITA	BRHR
		2.60	2.00	2.00	2.00					2.00

The Computational Science Graduate Fellowship program supports over 50 doctoral students in computational science and engineering.

Agency Ties	
ARPA	
NSF	
DOE	
NASA	
NIH	
NSA	
NIST	
NOAA	
EPA	
ED	

FY 1993 Accomplishments	FY 1994 Major Milestones	FY 1995 Major Milestones	Milestone Changes from Prior IP
Conduct first conference for all computational science fellows with uniformly excellent presentations.	Maintain current numbers of fellows. Conduct second workshop for fellows.	Evaluate quality of first group of fellows as they enter the professional computational science career path.	

DOE	Budget Account Code	Budget (BA, $ M)				HPCC Components by 1995 Pres. Req.				
Education, Training and Curriculum		FY 93 Actual	FY 94 Pres.	FY 94 Est.	FY 95 Pres.	HPCS	NREN	ASTA	IITA	BRHR
		2.36	3.00	3.00	3.00					3.00

Support for kindergarten through graduate school educational programs. Adventures in Supercomputing (AiS) trains in-service teachers in the use of computers and networks. AiS also encourages the participation of women and minorities in computational science. The National High School Honors Program provides summer enrichment in supercomputing for gifted high school students. Textbook projects develop instructional materials for undergraduate and graduate courses in computational science. A curriculum enhancement project is developing approaches for effecting permanent changes in university curricula at both the undergraduate and graduate levels. This activity also addresses the issue of engaging university administrators in the process of advocating computational science and engineering education.

Agency Ties	
ARPA	
NSF	
DOE	
NASA	
NIH	
NSA	
NIST	
NOAA	
EPA	
ED	

FY 1993 Accomplishments	FY 1994 Major Milestones	FY 1995 Major Milestones	Milestone Changes from Prior IP
AiS program established in New Mexico and Tennessee. Release beta version of graduate computational science electronic textbook.	Hold the first K-12 Administrators Workshop. Expand AiS program from New Mexico and Tennessee to Alabama, Colorado, and Iowa. Initiate undergraduate textbook project.	Make second general release of electronic computational science text Continue AiS program at current level, which is to introduce approximately 50 teachers and 25 school districts into the program. Continue development of undergraduate electronic computational science textbook Hold workshop on the role of schools of education in computational science education	

DOE	Budget Account Code	Budget (BA, $ M)				HPCC Components by 1995 Pres. Req.				
		FY 93 Actual	FY 94 Pres.	FY 94 Est.	FY 95 Pres.	HPCS	NREN	ASTA	IITA	BRHR
Basic Research for Applied Mathematics Research		15.25	15.70	15.70	16.00					16.00

This program activity supports applied mathematics, computer science, and computational science research that develops tools for enabling solutions to large scientific problems. DOE supports leading-edge research at 10 DOE laboratories and over 30 universities. Current topics include:

- numerical analysis and scientific computing,
- modeling and analysis of physical systems,
- dynamical systems theory and chaos,
- geometric and symbolic computation, and
- optimization theory and mathematical programming.

Funding also covers conferences and workshops in applied mathematics and computational science, the activities of the National Academy of Science's Board on Mathematical Sciences, over 40 post-doctoral associates at laboratories and universities, and over 60 university graduate students. Applied mathematicians and computational scientists supported by basic research are also active in Grand Challenge projects. The applied mathematics program was reviewed in FY 1993 by panels of outside experts. It received uniformly high marks for scientific excellence and relevance to the DOE mission. The panel also identified new areas for possible funding (inverse scattering, nondestructive evaluation, discrete mathematics, and spatio-temporal chaos).

Agency Ties	
ARPA	
NSF	Partner
DOE	
NASA	
NIH	
NSA	
NIST	
NOAA	
EPA	
ED	

FY 1993 Accomplishments	FY 1994 Major Milestones	FY 1995 Major Milestones	Milestone Changes from Prior IP
Novel applications of asymptotic analysis led to the development of improved device simulation codes for designing silicon and GaAs semiconductors. Analytical and numerical methods for large-scale stochastic programming problems provided the first successful solution of realistic linear and nonlinear programming problems in the presence of uncertainty. Computational geometers settled a century- old conjecture on the existence of certain types of minimal surfaces using a combined analytical/ computational/ visualization approach. A fast multigrid solver for transport problems was incorporated into DOE laboratory production codes enabling faster and more accurate solution of neutron scattering problems.	Make available a code that can resolve fluid dynamic instabilities produced by shock waves to the extent that simulation results agree with experiments in one and two dimensions. Design and implement numerical algorithms for controlling in real time, the behavior of chaotic responses in low-dimensional mechanical and electrical systems. Develop software and computer networking tools to enable chemists and material scientists to visualize complex geometric structures found in commercially important natural and synthetic materials.	Make available codes based on adaptive mesh refinement techniques for resolving 3D fluid flow phenomena in complex geometries. Design security procedures for voice and data transmission based on sending the data encapsulated in a chaotic signal and decoding the signal using techniques for controlling chaos. Make available codes for resolving large-scale stochastic programming problems of the type found in large commercial electric power distribution systems, data transmission networks, and airline crew equipment scheduling operations.	An environmental remediation modeling project at Oak Ridge National Laboratory was cancelled following a recommendation from the 1993 Program Review Panel that the researchers lacked the appropriate physical and numerical expertise to model realistic groundwater contaminant transport phenomena. Two applied mathematics projects at Sandia National Laboratory were cancelled. In one project on combustion modeling the researcher had perfected his bifurcation analysis techniques to the point where further support was deemed unnecessary, while in the second project on developing adaptive numerical schemes the researcher became involved in another project and was unable to continue working on this one.

National Aeronautics and Space Administration

NASA's primary role in this program includes leading the development of applications software and algorithms for scalable parallel computing systems which will increase system performance to the sustained teraFLOPS (1012 floating point operations per second) level for NASA applications. As HPCC technologies are developed, NASA will use them to solve its Grand Challenge research problems. These are fundamental problems whose solutions require significant increases in computational power and are critical to meeting national needs. NASA's Grand Challenges include improving the design and simulation of advanced aerospace vehicles, enabling people at remote locations to communicate more effectively and share information, increasing scientists' abilities to model the Earth's climate and forecast global environmental trends, and improving the capabilities of advanced spacecraft to explore the Earth and solar system.

NASA has initiated a program component to broaden the reach of the HPCC program and begin the development of a National Information Infrastructure by supporting research and development in education, digital library technology, and access to Earth and space science data. The FY 1995 NASA HPCC program has been organized into three vertically integrated computing projects: Computational Aerosciences (CAS) project, the Earth and Space Sciences (ESS) project, the Remote Exploration and Experimentation (REE) project and the Information Infrastructure Technology and Applications (IITA) component. The REE project will not be active from FY 1993 through FY 1995, but is expected to resume activities in FY 1996. In cooperation with other participating federal agencies, NASA's HPCS component involves deployment of experimental, scalable supercomputer capabilities essential for computational design of integrated aerospace vehicle systems and for predicting long-term global changes. NASA leads the coordination of the ASTA element among government, academic and industry partners. This includes conducting a series of workshops which address federal HPCC program needs in software technologies and algorithms, computational aerosciences and supercomputing testbeds. NASA also is responsible for implementing the Software Sharing Experiment to foster software sharing and reuse across the federal HPCC program. NASA also is acquiring hardware for computational science testbeds and developing software tools to enhance productivity, including load balancing tools, run time optimizers, monitors, parallelization tools as well as data management and visualization tools. The agency's NREN activity provides high-speed network connections among NASA, industry and academic researchers. NASA's Basic Research and Human Resources component encourages research into the underlying theory and concepts of high performance computing and will foster research in high performance computing at NASA centers, research institutes and universities. The new IITA component consists of critical information technologies and the application of these technologies to "National Challenges," problems where the application of HPCC technology can provide huge benefits to all Americans.

National Aeronautics and Space Administration

Program Activity	Budget Account Code	Partner/User Agencies	Budget (BA, $ M) FY 93 Actual	FY 94 Pres.	FY 94 Est.	FY 95 Pres.	HPCC Components by 1995 Pres. Req. HPCS	NREN	ASTA	IITA	BRHR
Testbeds	509-10 509-20 509-30	ARPA, NSF, DOE	17.60	26.30	22.60	26.40	9.70		16.70		
Grand Challenge Support	509-10 509-20	NSF, DOE, NOAA, EPA	46.80	57.40	53.80	55.30			55.30		
Systems Software	509-10 509-20 509-30	EPA, NOAA, NSA, NIH, ...	5.40	8.20	8.20	9.20			9.20		
NREN	509-10 509-20	ED, NOAA, EPA, NSF, ...	8.50	14.20	13.10	12.70		12.70			
BRHR	509-10 509-20	NSF, NIH, NOAA, EPA	3.30	3.40	3.30	3.80					3.80
Information Infrastructure Technology	509-40	DOE, NIH, ED, EPA, NOAA,	0.00	4.00	3.20	6.80				6.80	
Information Infrastructure Applications	509-40	NOAA, EPA	0.00	8.00	8.80	10.70				10.70	
		Totals:	81.60	121.50	113.00	124.90	9.70	12.70	81.20	17.50	3.80

NASA	Budget Account Code	Budget (BA, $ M)				HPCC Components by 1995 Pres. Req.				
		FY 93 Actual	FY 94 Pres.	FY 94 Est.	FY 95 Pres.	HPCS	NREN	ASTA	IITA	BRHR
Testbeds		17.60	26.30	22.60	26.40	9.70		16.70		

The objective of this element is to encourage and accelerate U.S. commercial development of high-performance computing systems that will support the Grand Challenges. To accomplish this, NASA will fund and encourage the commitment to early acquisition, access and placement of such advanced systems by:

- Acquiring some advanced prototype and early production model high-performance computing systems for use and evaluation;
- Providing network access to understand execution of full Grand Challenge applications on TeraFLOPS systems;
- Providing a testbed control environment to assist in the collection of data about testbed operations which are needed to project the eventual performance of Grand Challenge applications on TeraFLOPS computing systems;
- Developing a set of parallel benchmark codes based directly on the Grand Challenge applications to evaluate disparate architectures.

To compare different approaches to TeraFLOPS systems on a common basis, NASA will develop these parallel benchmarks to reflect the computational demands of the various Grand Challenge areas. All benchmarks will be scalable and used on HPCC testbeds.

NASA also will develop tools and techniques to project the execution performance of the Grand Challenge applications. High-performance computing research facilities will be established to accelerate transition to new generations of high-performance computing technology by enabling researchers to explore applications of these new technologies. This will provide early access to advanced computing facilities for advanced application experiments and facilitate the creation and evaluation of new computing technology. Advanced high-performance processors will be integrated into a prototype TeraFLOPS computing facility that will include access to the National Research and Education Network (NREN), early systems or advanced prototypes of important storage hierarchy subsystems, and sufficient advanced visualization facilities to allow system scalability experiments to be conducted. This will provide a spectrum of experiments for scalability studies as well as access to the largest possible advanced system by the Grand Challenge researchers.

Agency Ties	
ARPA	Partner
NSF	Partner
DOE	User
NASA	User
NIH	
NSA	
NIST	
NOAA	·
EPA	
ED	

FY 1993 Accomplishments	FY 1994 Major Milestones	FY 1995 Major Milestones	Milestone Changes from Prior IP
NASA played an active role in the Concurrent SuperComputer Consortium (CSCC) during its first year of operation. Use of the Consortium's Intel Touchstone Delta supercomputer has enabled major advances in several scientific and engineering applications including the processing of three-dimensional images of Venus taken by the Magellan satellite at the rate of four frames per second -- thus putting real-time image processing within reach. Other accomplishments on the Delta include the largest direct numerical simulation of the time-dependent compressible Navier-Stokes equations and the largest three-dimensional compressible turbulence simulations for high Reynolds numbers.	Install 10-50 GigaFLOPS sustained testbeds (scalable to 100 GigaFLOPS) for Computational Aerosciences (CAS) Grand Challenge teams. Foster teaming of independent software vendors, aerospace industry and computer manufacturers through access to 10-50 GigaFLOPS testbed.	Develop multidisciplinary aeroscience and Earth and Space Science (ESS) applications on 10-50 GigaFLOPS testbeds.	

NASA	Budget Account Code	Budget (BA, $ M)				HPCC Components by 1995 Pres. Req.				
		FY 93 Actual	FY 94 Pres.	FY 94 Est.	FY 95 Pres.	HPCS	NREN	ASTA	IITA	BRHR
Grand Challenge Support		46.80	57.40	53.80	55.30			55.30		

This research area develops, enhances and evaluates techniques for the multidisciplinary modeling and simulation required by Grand Challenge problems. The Computational Aeroscience (CAS) research will focus on understanding the high-performance computing environment and how it can be used to solve a range of problems of importance in aerospace engineering at a cost that represents the value, flexibility and short cycle time required by the aerospace community. This will be accomplished with the aid and focus provided by several representative aerospace design problems related to nationally important programs. These are: a High Speed Civil Transport (HSCT); an Advanced Subsonics Civil Transport (ASCT); High-Performance Aircraft (HPA), and Rotorcraft.

Earth and Space Science (ESS) research will cover two critical scientific areas: the coupling of advanced discipline models into scalable global simulations providing realistic global change understanding; and the integration of models and analysis algorithms for processing, analyzing and understanding the enormous volumes of data expected from scientific missions. Research will focus on: large scale structure and galaxy formation; cosmology and accretion astrophysics; convective turbulence and mixing in astrophysics; solar activity and heliospheric dynamics; Earth system models; four-dimensional data assimilation; climate models, and knowledge discovery in geophysical databases and satellite data.

Collaborative groups including discipline scientists, software and systems engineers, professional software developers and algorithm designers are supported by shared computational and experimental facilities. Technical accomplishments will include development of application-specific codes for innovative high-performance computing systems, design and analysis of algorithms, and architecture and performance assessment of specific applications. NASA will evaluate the early research products and make the results available to system vendors as quickly as possible. Results in design and theory of algorithms are as important to breaking down computational scaling barriers as are performance improvements in computing hardware. Algorithms for common techniques such as multidimensional FFTs, Fast Poisson solvers, multigrid methods, Reimann solvers, sparse matrix methods, singular value decomposition, matrix factorization methods, and spectral methods are being re-implemented on a variety of architectures in order to understand how architecture affects efficiency and algorithm design.

Agency Ties	
ARPA	
NSF	Partner
DOE	Partner
NASA	
NIH	
NSA	
NIST	
NOAA	User
EPA	User
ED	

FY 1993 Accomplishments	FY 1994 Major Milestones	FY 1995 Major Milestones	Milestone Changes from Prior IP
NASA established Grand Challenge applications interdisciplinary teams in CAS and ESS. The Concurrent SuperComputer Consortium (CSCC) enabled major advances in the processing of three-dimensional images of Venus taken by the Magellan satellite at the rate of four frames per second, the largest direct numerical simulation of the time-dependent compressible Navier-Stokes equations, and the largest 3-D compressible turbulence simulations for high Reynolds numbers. NASA ported single discipline computational fluid dynamics code and structural analyses to a number of scalable parallel computers. NASA developed and demonstrated the ENSAERO computer algorithm, which couples aerodynamics and structures analysis modules on the Intel iPSC/860 computer.	Aeronautics industry review and validation of NASA aeroscience computational tools to promote industrial use. The Computational Aerosciences Consortium of U.S. industrial firms has worked with NASA to facilitate the development of pre-competitive software for the implementation of integrated multidisciplinary design, analysis, and optimization systems on heterogeneous computer networks.	Develop multidisciplinary aeroscience and Earth and space science applications on 10-50 GigaFLOPS testbed: 5-10 GigaFLOPS sustained performance on Earth science applications. Architecture evaluation results by Earth and Space Science Team. Advanced subsonic civil transport analysis/optimization code; and medium fidelity high speed civil transport modeling. Enhance the ability of aircraft manufacturers to quickly analyze different design options and accelerate the prototyping process, thereby reducing design cycle costs and producing vehicles with improved performance.	

NASA	Budget Account Code	Budget (BA, $ M)				HPCC Components by 1995 Pres. Req.				
		FY 93 Actual	FY 94 Pres.	FY 94 Est.	FY 95 Pres.	HPCS	NREN	ASTA	IITA	BRHR
Systems Software		5.40	8.20	8.20	9.20			9.20		

There are common needs in many areas of software technology including programming environments for code development and adaptation, advanced compiler technology, tools for optimization and parallelization, data management and interoperability, analysis and performance measurements, user interaction and visualization, and debugging and instrumentation. Advances in these generic software technology areas will have broad national impact resulting from the united efforts of NASA, other Federal agencies and industry. As the lead agency for coordinating advanced software technology and algorithm development, NASA will ensure that new systems software developments will involve broad communities. NASA is leading the effort to develop a HPCC Software Exchange to provide the infrastructure that encourages software reuse and the sharing of software modules across organizations through an interconnected set of software repositories.

Research will be conducted in the development of program debugging tools and in instrumentation facilities for monitoring execution and developing new techniques for monitoring and presenting the state of concurrent program execution in a coherent and user-friendly manner. Studies will include evaluating the scalability of these utilities.

Research also will be conducted in the design of data management software needed to support the development and use of Grand Challenge based applications on future highly parallel systems. Techniques will be explored to control efficient, high-performance I/O in highly parallel computer systems. The structures which are used heavily in multidisciplinary design applications may be object-oriented. Dynamic resource management methods will be prototyped and evaluated. The portability of these methods to various high-performance systems will be studied.

Human interfaces will be developed to permit users to observe and manipulate the huge amounts of 3D temporal input and result data from the multidisciplinary simulation, analysis, and optimization processes in a manageable, coherent fashion, and to allow for the analysis of the discrete physics through visualization, manipulation, and comparisons with other experimental or computational data.

Agency Ties	
ARPA	Partner
NSF	Partner
DOE	Partner
NASA	
NIH	User
NSA	User
NIST	
NOAA	User
EPA	User
ED	

FY 1993 Accomplishments	FY 1994 Major Milestones	FY 1995 Major Milestones	Milestone Changes from Prior IP
The HPCC Software Exchange Experiment encourages the sharing and reuse of software modules by providing an infrastructure of interconnected software repositories. It represents the first time that a number of different agencies with different philosophies for software management have collaborated to develop an integrated, albeit experimental, system. NASA also sponsored a number of national conferences to stimulate discussion among HPCC systems software and tool researchers and Grand Challenge applications teams regarding the development of a national research agenda for HPCC systems and applications software and to promote coordination among agency-sponsored efforts. These efforts are leading to teaming arrangements between research labs in different federal agencies.	Organize second workshop on systems software and tools for High Performance computing environments.	Improve portability and scalability of software components and tools to high performance computing systems.	

NASA	Budget Account Code	Budget (BA, $ M)				HPCC Components by 1995 Pres. Req.				
		FY 93 Actual	FY 94 Pres	FY 94 Est	FY 95 Pres.	HPCS	NREN	ASTA	IITA	BRHR
NREN		8.50	14.20	13.10	12.70		12.70			

The National Research and Education Network (NREN) effort will establish standards and provide working models for commercial communications infrastructure deployment. NASA's role is to deploy the advanced communications required by the Grand Challenge investigators in a manner that satisfies the immediate needs of researchers while simultaneously guiding commercial infrastructure development for the nation. NASA works with NSF, DOE, ARPA, and other agencies to enhance the national network infrastructure by coordinating the development and implementation of enhanced network technologies and services: integrated voice, video, and computer data transmission; network management and operations tools; protocol standards; routers and switches; security management; emerging high-performance user services including provision of advanced network services for multi-media communications; and K-12 distance learning and outreach applications.

NASA has already deployed 45 Mbps network services between two NASA centers and the NSFnet, significantly improving communications between remote investigators and NASA. An acquisition for advanced telecommunications services on an early availability basis will bring telecommunication and computational standards together to provide a low-cost computer network infrastructure over vendor facilities ultimately targeted at commercial availability. This will take advantage of the latest telecommunications technologies, such as Asynchronous Transfer Mode (ATM) over Synchronous Optical Network Transmission (SONET) services, initially at 45 Mbps among the five NASA centers.

NASA also is working with other Federal HPCC agencies to extend the reach of NASA's data resources and computing capabilities to K-12 schools, libraries, and teacher resource centers. Further development of digital libraries and other networking infrastructure technologies under the IITA component will promote remote access to NASA HPCC resources, enabling schools and resource centers to collaborate directly with each other over networks on distributed NASA education projects.

In order to incorporate satellite-based communications technology into the terrestrial gigabit highways of the future, two experiments are planned for coupling the NREN effort with the NASA's Advanced Communications Technology Satellite (ACTS).

Agency Ties	
ARPA	Partner
NSF	Partner
DOE	Partner
NASA	
NIH	
NSA	
NIST	
NOAA	User
EPA	User
ED	User

FY 1993 Accomplishments	FY 1994 Major Milestones	FY 1995 Major Milestones	Milestone Changes from Prior IP
NASA supports the architecture evolving in the NREN component and participates in the development and deployment of gigabit network technologies and architectures. NASA continues to develop and deploy advanced networking technologies to allow researchers and educators to carry out collaborative research and educational activities. NASA has entered into a cooperative effort with DOE to procure network services that will operate at 45 Mbps using synchronous optical fiber network standards. The two agencies are working closely to provide a nationwide fiber optic network that will meet HPCC research communications needs and serve as the foundation for even faster networks ranging from 155 Mbps to eventual gigabits speeds.	Complete T-3 (45 Mbps) Level 3 HPCC interconnects of five NASA centers -- Ames Research Center (ARC), Goddard Space Flight Center (GSFC), Langley Research Center (LaRC), Lewis Research Center (LeRC) and the Jet Propulsion Laboratory (JPL) -- with switched packet network to serve research communities.	Demonstrate satellite-based Gigabit applications using the Advanced Communications Technology Satellite (ACTS) including: "A Performance Study of a SONET/ATM Satellite & Terrestrial Network for an Engine Inlet Simulation," and "High Data ACTS Experiments for Performing Global Science: Keck Telescope and Global Climate Model." Interconnect five NASA research centers using 155 Mbps high speed communication: Ames Research Center (ARC), Goddard Space Flight Center (GSFC), Langley Research Center (LaRC), Lewis Research Center (LeRC) and the Jet Propulsion Laboratory (JPL).	

NASA	Budget Account Code	Budget (BA, $ M)				HPCC Components by 1995 Pres. Req.				
		FY 93 Actual	FY 94 Pres.	FY 94 Est.	FY 95 Pres.	HPCS	NREN	ASTA	IITA	BRHR
BRHR		3.30	3.40	3.30	3.80					3.80

Effective integration of new high-performance computing technology into the U.S. mainstream will require a sustained research effort across the spectrum of computing technology. Areas to be included are: computer architecture; fundamental algorithms; computational complexity; networked and distributed computation; numerical analysis, and application specific algorithms.

NASA supports research institutes and centers of excellence engaged in computer science and computational science under the basic research and human resources component. These include: the Illinois Computer Laboratory for Aerospace Systems and Software (ICLASS) at the University of Illinois; the Research Institute for Advanced Computer Science (RIACS) at Ames Research Center; the Institute for Computer Applications in Science and Engineering (ICASE) at Langley Research Center, and the Center of Excellence for Space Data and Information Sciences (CESDIS) at Goddard Space Flight Center. In addition, NASA funds small university grants on HPCC topics at the individual principal investigator level.

The NASA HPCC program also focuses on the development of the next generations of computer and computational scientists. The resource potential to be developed will be concentrated at the graduate and post-doctoral level, but will extend to the baccalaureate and junior professor degree levels to a lesser extent. NASA's HPCC program will address the need to attract more talented scientists and engineers to computer and computational science in multiple ways by:

- Expanding ongoing NASA programs which support pre-baccalaureate and master's level training and research programs (e.g., ICLASS);
- Expanding doctoral degree research opportunities in High-Performance Computing (e.g., CESDIS, ICLASS);
- Expanding the ongoing NASA programs which support post-doctoral research and new professors (e.g., CESDIS, ICASE, RIACS, ICOMP).

In addition, new mechanisms for supporting students and new faculty interested in applying HPCC technology on NASA's applications have been initiated. This includes funding students directly at their institutions provided they have an advisor interested in NASA applications and expanding on the NASA Graduate Student Researchers Program at NASA centers.

Agency Ties	
ARPA	
NSF	Partner
DOE	
NASA	
NIH	User
NSA	
NIST	
NOAA	User
EPA	User
ED	

FY 1993 Accomplishments	FY 1994 Major Milestones	FY 1995 Major Milestones	Milestone Changes from Prior IP
Pilot programs with the K-12 educational community have been established to discover the best mechanisms for using space and aeronautics assets to support educational communities in the U.S., and to provide practical models for using sophisticated computational and networking resources in these communities. This was and will continue to be accomplished in collaboration and Grand Challenge scientists from the CAS and ESS projects, thereby achieving direct involvement between NASA scientists and the K-12 community.	Provide annual graduate & post-doctoral support for high performance computing research.	Provide annual graduate & post-doctoral support for high performance computing research.	

NASA	Budget Account Code	Budget (BA, $ M)				HPCC Components by 1995 Pres. Req.				
		FY 93 Actual	FY 94 Pres.	FY 94 Est.	FY 95 Pres.	HPCS	NREN	ASTA	IITA	BRHR
Information Infrastructure Technology		0.00	4.00	3.20	6.80				6.80	

NASA will promote the development and deployment of digital libraries within the National Information Infrastructure in partnership with other federal agencies as called for in the administration's technology plan and pending legislation in the House (H.R. 1757) and Senate (S.4). NASA will follow a three fold approach in pursuing this project based on both NASA unique requirements and technology contributions and on more general national requirements for advanced technology for digital libraries.

First, in conjunction with the NII Applications of Remote Sensing Images and Data project, NASA will issue a Cooperative Agreement Notice for research and development of digital library technology to support the access to and distribution of remote sensing images and data over computer networks such as the Internet. The proposals selected under this project will primarily be for cooperative research efforts to develop both innovative and scalable applications of these images and data in both the public and private sector. Communities and applications that will potentially be served include: agriculture, forestry, state and local governments, transportation, education, and traditional library services.

The second part of NASA's program in digital library technology will be to develop more generic technologies for the longer term. This will be done in partnership with the National Science Foundation and the Advanced Research Projects Agency to economize the procurement of this crosscutting research. A joint NSF/ARPA/NASA solicitation is in preparation for release in FY '94.

The final component of NASA's Digital Library Technology project will be to foster the development of standards for library interoperability. Since ultimately digital library technologies will be developed and supported by the private sector, early involvement by the private sector is crucial to expeditiously develop standards. NASA's strategy for this area will be to promote a joint government, industry and academia Digital Library Technology Team to develop a standards development process and to initiate the standardization of key technologies to accelerate the creation of NII Digital Libraries.

Agency Ties	
ARPA	Partner
NSF	Partner
DOE	User
NASA	
NIH	User
NSA	
NIST	
NOAA	User
EPA	User
ED	User

FY 1993 Accomplishments	FY 1994 Major Milestones	FY 1995 Major Milestones	Milestone Changes from Prior IP
Prior to element start.	Initiate cooperative agreements with the non-federal sector for the development of digital library technology.	Establish a Remote Sensing Public Access Center for serving the Internet communities' requirements with regard to accessing NASA Earth and space science data.	

NASA	Budget Account Code	Budget (BA, $ M)				HPCC Components by 1995 Pres. Req.				
		FY 93 Actual	FY 94 Pres.	FY 94 Est.	FY 95 Pres.	HPCS	NREN	ASTA	IITA	BRHR
Information Infrastructure Applications		0.00	8.00	8.80	10.70				10.70	

NASA will develop and provide access to databases of remote sensing images and supportive software over computer networks like the Internet. Such databases would be accessible to both public and private institutions promoting the dissemination of taxpayer-funded Federal information. The information contained within these databases can also be used by the educational and library communities to fulfill their needs and goals. The programmatic goal is to provide broad public access to remote sensing data to traditionally un-served communities. This data can be NASA and other federally funded data provided to the application communities. Providing the access to remote sensing data will promote gains in education, quality of life and economic growth.

NASA will build on its HPCC program, its aeronautics and space science research and engineering missions, and its existing education outreach infrastructure to facilitate the general development of the National Information Infrastructure to support mathematics, science, and engineering education in the K-12 levels. In FY '94, NASA will issue a research announcement or cooperative agreement notice open to public and private parties nationwide. The intent of this announcement will be to solicit and fund proposals pursuing innovative ways to integrate the emerging national information infrastructure (the National Research and Education Network, open repositories of math and science information, state of practice networking tools like WAIS, Gopher, and Mosaic, ...) into established and accepted educational curricula.

Agency Ties	
ARPA	
NSF	
DOE	
NASA	
NIH	
NSA	
NIST	
NOAA	Partner
EPA	Partner
ED	

FY 1993 Accomplishments	FY 1994 Major Milestones	FY 1995 Major Milestones	Milestone Changes from Prior IP
Prior to element start.	Initiate first set of cooperative agreements for innovative applications of remote sensing databases.	Demonstrate applications for public access and retrieval of NASA databases over the Internet for commercial and public purposes by traditionally under-served communities (i.e. schools, public libraries, small businesses and state or local governments). Demonstrate Internet access and computational science curriculum tools for K-12 education including: NASA "Daytime Astronomy" tools from Project SEED (Science in Early Education Development); Information search and discovery tool for teachers (ILIAD) for use over the Internet; and Internet training for the Science Teacher Ambassadors (representing 24 school districts).	

National Institutes of Health

The National Institutes of Health (NIH) contribution to the High Performance Computing and Communications initiative focusses on biomedical applications of computing and digital communications. HPCC efforts include research and development within NIH's intramural laboratories, and sponsorship via grants and contracts of HPCC efforts at educational and research institutions. The program includes Grand Challenges research in areas such as biomolecular sequence and structure analysis, medical image reconstruction and processing, and the creation and evaluation of advanced computing architectures in these domains. The IITA component of the NIH program focusses on healthcare-related activities, such as the development of testbed networks linking hospitals, clinics, libraries, and medical schools, and the development of computerized patient records and telemedicine technologies. HPCC programs are administered by the National Library of Medicine (NLM), National Center for Research Resources (NCRR), Division of Computer Research and Technology (DCRT) and the National Cancer Institute (NCI).

National Institutes of Health

Program Activity	Budget Account Code	Partner/User Agencies	Budget (BA, $ M)				HPCC Components by 1995 Pres. Req.				
			FY 93 Actual	FY 94 Pres.	FY 94 Est.	FY 95 Pres.	HPCS	NREN	ASTA	IITA	BRHR
NLM Medical Connections Program		NSF	0.40	3.70	0.60	6.50		6.50			
NLM Biotechnology Informatics			4.10	9.50	4.10	4.80			4.80		
NLM Electronic Imaging			1.50	4.10	2.20	2.20			2.20		
NLM HPCC Training Grants			2.90	3.60	2.90	3.60					3.60
NLM IAIMS grants			3.80	7.60	3.80	5.40					5.40
NLM Intelligent Agent DB searching			5.00	4.40	5.60	4.70				4.70	
NLM HPCC Health Care Applications			1.50	6.80	7.00	11.00				11.00	
NCRR Advanced Software Technology and Algorithms		NSF	6.80	8.80	8.80	8.80			8.80		
NCRR Information Infrastructure Technology Applications		NSF	3.40	4.70	4.70	8.80				8.80	
NCRR Basic Research and Human Resources		NSF	3.10	2.30	2.30	5.00					5.00
DCRT High Performance Biomedical Computing Program		ARPA	8.00	8.40	8.40	11.00	4.00	1.00	3.00	3.00	
NCI Frederick Biomedical Supercomputing Center		NSA, NIST	6.30	6.70	6.70	6.70	0.90	0.40	4.60		0.80
NCI High speed networking and distributed conferencing						2.00		0.50	0.40	1.00	0.10
High Perf. Communications for PDQ, CancerNet, and Electronic Publishing						0.60				0.60	
National Coordination Office			0.40	0.90	0.70	0.70					0.70
		Totals:	47.20	71.50	57.80	81.80	4.90	8.40	23.80	29.10	15.60

NIH	Budget Account Code	Budget (BA, $ M)				HPCC Components by 1995 Pres. Req.				
		FY 93 Actual	FY 94 Pres.	FY 94 Est.	FY 95 Pres.	HPCS	NREN	ASTA	IITA	BRHR
NLM Medical Connections Program		0.40	3.70	0.60	6.50		6.50			

The Medical Connections grant program provides "jump start" funding to academic medical centers and health care organizations to allow them to connect to the NREN. Joint funding by the National Library of Medicine and the National Science Foundation is provided to offset the costs of digital communications equipment, digital circuits linking medical centers with academic and commercial mid-level networks, and personnel and services necessary to connect to the NREN.

Special emphasis is given to linking medical libraries with health care delivery organizations so that high speed telecommunications can support delivery of timely and accurate information for clinical decision making. The program also supports creation of regional consortia of health care institutions for sharing of medical information and data.

Agency Ties	
ARPA	
NSF	Partner
DOE	
NASA	
NIH	
NSA	
NIST	
NOAA	
EPA	
ED	

FY 1993 Accomplishments	FY 1994 Major Milestones	FY 1995 Major Milestones	Milestone Changes from Prior IP
15 Connections grants were awarded in 1993.	Approximately 20-25 Connections grants will be awarded in 1994	At full funding levels, approximately 100 grants would be awarded annually	Due to funding limitations, 1994 grant targets were reduced from 100 to 20-25.

NIH	Budget Account Code	Budget (BA, $ M)				HPCC Components by 1995 Pres. Req.				
NLM Biotechnology Informatics		FY 93 Actual	FY 94 Pres.	FY 94 Est.	FY 95 Pres.	HPCS	NREN	ASTA	IITA	BRHR
		4.10	9.50	4.10	4.80			4.80		

The NLM's National Center for Biotechnology Information (NCBI) has been given the legislative mandate to create automated systems for storing and analyzing the vast and growing ocean of data related to molecular biology, biochemistry, and genetics. This information science is an essential component of genome research and protein engineering/drug design, and seeks to develop analytical and predictive methods to identify key molecular patterns associated with health and disease as represented in a large number of related databanks encoding DNA, RNA, proteins, and other biologically important molecules.

NCBI is implementing a "backbone" sequence database of molecular sequence and structure which contains key linkages to the scientific literature and to existing biological databanks; the database is available to researchers nationally via the NREN, using several advanced client-server prototype retrieval systems. Research on efficient and expressive data representation techniques for molecular sequence objects is a critical component of this effort, which complements basic research on the development of new sequence analysis and retrieval methods. Within this distributed database architecture, the Center builds and provides access to GenBank, the national DNA sequence databank which is a key data resource of the Human Genome Project.

The Biotechnology Informatics program also supports investigator- initiated research in computational biology via peer-reviewed grants.

Agency Ties	
ARPA	
NSF	
DOE	
NASA	
NIH	
NSA	
NIST	
NOAA	
EPA	
ED	

FY 1993 Accomplishments	FY 1994 Major Milestones	FY 1995 Major Milestones	Milestone Changes from Prior IP
Creation and public distribution of GenBank, the national DNA sequence databank, in a format integrated with other molecular sequence databanks and the biomedical literature of Medline. Public availability of "Entrez" integrated database search engine.	Creation of automated direct author submission methods with automated quality control analysis algorithms. 2-5 investigator-initiated grants in biotechnology algorithm development.	Tenfold expansion of available gene sequences linked with protein and literature references.	

NIH	Budget Account Code	Budget (BA, $ M)				HPCC Components by 1995 Pres. Req.				
		FY 93 Actual	FY 94 Pres.	FY 94 Est.	FY 95 Pres.	HPCS	NREN	ASTA	IITA	BRHR
NLM Electronic Imaging		1.50	4.10	2.20	2.20			2.20		

Images are an important part of biomedical knowledge. New computer-based technologies are providing an unprecedented opportunity to supplement the traditional two dimensional images of medicine and biology with dynamic, three-dimensional images that can be viewed, rotated, and reversibly dissected in a manner analogous to the physical objects they represent. The National Library of Medicine has undertaken the first steps to build digital image libraries of the three dimensional structure of the human body at sub-millimeter level resolution. Full use and understanding of the biological structure depicted in such libraries will require advanced computing methods to identify and classify biological objects within image sets, as well as algorithms to efficiently render anatomic detail acquired from computed tomography (CT), magnetic resonance imaging (MRI), and anatomical prosection in a manner that is photo realistic and easily manipulable by the student, researcher, or health care provider. The NLM is also working with industry and academic groups to develop standards for the representation and communication of such electronic images.

Agency Ties	
ARPA	
NSF	
DOE	
NASA	
NIH	
NSA	
NIST	
NOAA	
EPA	
ED	

FY 1993 Accomplishments	FY 1994 Major Milestones	FY 1995 Major Milestones	Milestone Changes from Prior IP
Data acquisition completed for a single cadaver at millimeter level resolution.	Complete sample-based datasets for representative normal human anatomy will be made available by NLM via the Internet/NREN.	First object classification and geometry based image libraries available.	

NIH	Budget Account Code	Budget (BA, $ M)				HPCC Components by 1995 Pres. Req.				
		FY 93 Actual	FY 94 Pres.	FY 94 Est.	FY 95 Pres.	HPCS	NREN	ASTA	IITA	BRHR
NLM HPCC Training Grants		2.90	3.60	2.90	3.60					3.60

Currently, there are not adequate numbers of persons in biomedical fields who have had training in the use of modern computer and communications systems. There is a need both for biomedical professionals cross-trained in informatics and for persons from computer and information sciences and engineering who have had doctoral or post-doctoral training in the application of these technologies to health problems. The major university medical centers that have concentrated their resources on planning for institution-wide information services have found that locating and recruiting senior professionals with this kind of education and training is their greatest obstacle. Training in health information management skills is critical. NLM's is expanding its successful pre- doctoral and post-doctoral grants program for career training in medical informatics, and providing an HPCC-in-medicine fellowship training program.

Agency Ties	
ARPA	
NSF	
DOE	
NASA	
NIH	
NSA	
NIST	
NOAA	
EPA	
ED	

FY 1993 Accomplishments	FY 1994 Major Milestones	FY 1995 Major Milestones	Milestone Changes from Prior IP
10 institutional training grants awarded.	Continuing support for existing programs continued.	Additional 1-2 institutional training grants to be awarded.	Number of awarded not increased due to funding limitations.

NIH	Budget Account Code	Budget (BA, $ M)				HPCC Components by 1995 Pres. Req.				
		FY 93 Actual	FY 94 Pres.	FY 94 Est.	FY 95 Pres.	HPCS	NREN	ASTA	IITA	BRHR
NLM IAIMS grants		3.80	7.60	3.80	5.40					5.40

Academic Medical Centers are the backbone of the American biomedical research enterprise. These 120 institutions are comprised of medical schools with their associated teaching and research hospitals, clinics and laboratories. Research-related information is the life blood of these complex centers, and increasingly this research information is in electronic form: databases of bibliographic and factual information, molecular databases, patient records, laboratory and clinical data. Currently, these electronic sources are largely disconnected and isolated from one another, and communications between computerized systems in academic medical centers is primitive or non-existent. The focus of the IAIMS program, initiated in 1984, is the development of advanced algorithms to link and retrieve conceptually related information from many disparate sources within the medical center, and between medical centers which are distant from one another. The goal of the program is the development, testing, and implementation of generalizable models of information flow management within university health science centers in a phased approach. The expected outcomes of this program are improved research productivity of publicly-supported research investigators at academic centers, improved access to patient data for technology assessment and health outcomes research, and more efficient patient care leading to reductions in unnecessary hospitalization.

Agency Ties	
ARPA	
NSF	
DOE	
NASA	
NIH	
NSA	
NIST	
NOAA	
EPA	
ED	

FY 1993 Accomplishments	FY 1994 Major Milestones	FY 1995 Major Milestones	Milestone Changes from Prior IP
Nine IAIMS grant and research contract awards were made to academic medical centers in seven states and the District of Columbia; four of these sites are implementing working systems which integrate academic information sources on local, regional, and national levels.	Thirteen IAIMS projects will be supported by grant and contract mechanisms, including two implementation continuation awards, one operations award and eight planning awards.	15-20 awards divided among institutional planning, implementation, and operations are anticipated; emphasis on sharing of information via the NREN will be a central focus of the program.	

NIH	Budget Account Code	Budget (BA, $ M)				HPCC Components by 1995 Pres. Req.				
		FY 93 Actual	FY 94 Pres.	FY 94 Est.	FY 95 Pres.	HPCS	NREN	ASTA	IITA	BRHR
NLM Intelligent Agent DB searching		5.00	4.40	5.60	4.70				4.70	

With the large and growing number of computerized data base resources and services offering bibliographic, full text and factual data bases, it is difficult for the user to locate and process needed information. One may not know where--in which data base-- to look, and the user must deal with the structured and unforgiving access protocols and retrieval languages that differ from one data base service to the next. An especially vexing problem arises if the user needs to search across several data bases or services containing information in multiple formats. Additionally, in biomedicine, the disparity in the biomedical terminology used to describe related concepts in different machine-readable files prevents practitioners and researchers from integrated relevant biomedical information from separate sources, such as the biomedical literature, clinical records, medical databanks, and expert knowledge bases. NLM's approach to these problems is to develop intelligent gateways among data base services, and a Unified Medical Language System (UMLS) to compensate for the dissimilarity in the ways related information is classified in different automated systems. Intelligent-agent-mediated gateways will give a user one point to request needed information and free the user as much as possible from having to know the peculiarities of the various information sources. The UMLS will function as an electronic Rosetta Stone, making the myriad of classifications of medical knowledge invisible to the user and enabling retrieval of related biomedical information from many sources.

Agency Ties	
ARPA	
NSF	
DOE	
NASA	
NIH	
NSA	
NIST	
NOAA	
EPA	
ED	

FY 1993 Accomplishments	FY 1994 Major Milestones	FY 1995 Major Milestones	Milestone Changes from Prior IP
Functioning Knowbot-mediated multi-database search engine demonstrated.	Scaling up of prototype engine to form Access gateway architecture.	Deployment and evaluation of Access gateway architecture for subset of NLM online services.	

NIH	Budget Account Code	Budget (BA, $ M)				HPCC Components by 1995 Pres. Req.				
		FY 93 Actual	FY 94 Pres.	FY 94 Est.	FY 95 Pres.	HPCS	NREN	ASTA	IITA	BRHR
NLM HPCC Health Care Applications		1.50	6.80	7.00	11.00				11.00	

Through a variety of mechanisms, the NLM is promoting the application of HPCC technologies to health care. The program supports research and development projects in each of the following IITA areas:

1. Testbed networks for linking hospitals, clinics, doctor's offices, medical schools, medical libraries, and universities to enable health care providers and researchers to share medical data and imagery.
2. Software and visualization technology for visualizing the human anatomy and analyzing imagery from X-rays, CAT scans, PET scans, and other diagnostic tools.
3. Virtual reality technology for simulating operations and other medical procedures.
4. Collaborative technology to allow several health care providers in remote locations to provide real-time treatment to patients.
5. Database technology to provide health care providers with access to relevant medical information and literature.
6. Database technology for storing, accessing, and transmitting patients' medical records while protecting the accuracy and privacy of those records.

Agency Ties	
ARPA	
NSF	
DOE	
NASA	
NIH	
NSA	
NIST	
NOAA	
EPA	
ED	

FY 1993 Accomplishments	FY 1994 Major Milestones	FY 1995 Major Milestones	Milestone Changes from Prior IP
A single award for an advanced statewide rural telemedicine networking project was made to a consortium led by the University of West Virginia.	Eleven additional contract awards will be made in the health care-related research areas of the BAA. NLM and AHCPR announced a joint cooperative agreement program for developing and testing computer-=based patient record systems. Several FY 1994 awards are planned. NLM will explore the use of consortia of federal agencies, federal laboratories, universities, and commercial companies to speed the development of health care applications of HPCC technologies.	Continuation funding of the twelve initial HPCC BAA three-year awards and the initial cooperative agreements. Fund five to ten new projects via contract grant or consortia mechanisms.	

NIH	Budget Account Code	Budget (BA, $ M)				HPCC Components by 1995 Pres. Req.				
		FY 93 Actual	FY 94 Pres.	FY 94 Est.	FY 95 Pres.	HPCS	NREN	ASTA	IITA	BRHR
NCRR Advanced Software Technology and Algorithms		6.80	8.80	8.80	8.80			8.80		

Advanced Software Technology and Algorithms (ASTA) activities in the National Center for Research Resources (NCRR) comprise algorithm and software development for advanced, high performance computing environments. The two major themes of this work are molecular biology and biomedical imaging. Molecular biology computing includes comparison of the similarities of genetic and protein sequence, and development of algorithms to predict molecular structure and function. Biomedical imaging includes imaging of molecules and new methods of correlating and displaying clinical images in three dimensions.

Agency Ties	
ARPA	
NSF	Partner
DOE	
NASA	
NIH	
NSA	
NIST	
NOAA	
EPA	
ED	

FY 1993 Accomplishments	FY 1994 Major Milestones	FY 1995 Major Milestones	Milestone Changes from Prior IP
Used 3-D simulation of a single heartbeat to better understand heart function; required 150 hours on Cray T3D time. Used simulation of an entire bio- logical membrane including all lipid and protein components to under- stand the mechanism of membrane function and regulation. Work done in conjunction with Eli Lilly. Developed new code to couple supercomputer systems with massively parallel processors for new molecular dynamics calculations. Developed new algorithms for registration and rendering of 3-D images from 2-D clinical images and micrographs. Designed new drugs to inhibit HIV. Determined influenza virus structure for new vaccines. Integrated graphics and molecular dynamics.	Complete development of new algorithms for protein folding simulations, calculation of free energy of solvation, potential evaluation, and sampling of conformational space. Implement efficient versions of several of the methods using the vector units on the CM-5 massively parallel supercomputer. Develop interface for viewing molecular dynamics and for influencing its course	Apply computer technology to accelerate protein structure determination using NMR data. Investigate the conformation of biologically important peptides. Investigate binding of drug molecules to receptors with known crystal structures. Apply integration of graphics and molecular dynamics to problems in protein design using the full power of a multi-processor supercomputer.	

NIH	Budget Account Code	Budget (BA, $ M)				HPCC Components by 1995 Pres. Req.				
		FY 93 Actual	FY 94 Pres.	FY 94 Est.	FY 95 Pres.	HPCS	NREN	ASTA	IITA	BRHR
NCRR Information Infrastructure Technology Applications		3.40	4.70	4.70	8.80				8.80	

Information Infrastructure Technology Applications (IITA) initiatives for the NCRR encompass important emerging high performance computing opportunities. Areas of emphasis are programs in simulation resources; image analysis; virtual reality technologies, group software development and science education. In the area of stimulation, major goals are to enhance computational capabilities and to project the course of epidemics, such as AIDS. Image analysis goals are real-time medical imaging and rational drug design. Virtual reality includes development of tools to expand the basic and clinical application of technological methods and instrumentation. Goals for group software development are to support group cooperative work environment and use of networks. Science education initiatives will emphasize the development of interactive learning tools to increase public and student understanding of science.

Agency Ties	
ARPA	
NSF	Partner
DOE	
NASA	
NIH	
NSA	
NIST	
NOAA	.
EPA	
ED	

FY 1993 Accomplishments	FY 1994 Major Milestones	FY 1995 Major Milestones	Milestone Changes from Prior IP
At the simulation resource for stochastic population models, software modules are being adapted to examine the spread of AIDS. Research focuses on the development and use of complex simulations of sociological structured populations. Develop a nano manipulator using a scanning tunneling microscope and a virtual environment interface. This technology could enable scientists to manipulate at the atomic level, opening the way for high perform- ance computer-assisted drug design. Pilot project to introduce new biomedical science computing technologies to high school science teachers and their students to enhance their classroom and laboratory lessons. Also includes guest speakers and seminars on educational computing.	Continue development of near-real-time algorithm and running code for the calculation of solvent- accessible surfaces around proteins and nucleic acids for docking studies and binding energy calculation. Use new high-resolution virtual reality system to display and study protein molecule structure. Using virtual reality technologies and environments to continue research and development on surgical procedures which would minimize surgical trauma to areas of the body that are difficult to access. Design and develop a national biomedical information "Kiosk" which would tie supercomputing, research data, and education together nationally.	Provide an inverse kinematics tool that should make it easier to determine possible configurations of protein chain segments with up to 12 degrees of freedom, such as turns connecting alpha helices to each other and to beta strands. Enable chemists to use new-technology SARCOS force display arm of high resolution accuracy and frequency response. This instrumentation is of particular importance in docking and folding studies.	

NIH	Budget Account Code	Budget (BA, $ M)				HPCC Components by 1995 Pres. Req.				
NCRR Basic Research and Human Resources		FY 93 Actual	FY 94 Pres.	FY 94 Est.	FY 95 Pres.	HPCS	NREN	ASTA	IITA	BRHR
		3.10	2.30	2.30	5.00					5.00

Basic Research and Human Resources (BRHR) NCRR initiative includes plans for cross-disciplinary training of established investigators in the use of advanced computing systems and methods.

Agency Ties	
ARPA	
NSF	Partner
DOE	
NASA	
NIH	
NSA	
NIST	
NOAA	
EPA	
ED	

FY 1993 Accomplishments	FY 1994 Major Milestones	FY 1995 Major Milestones	Milestone Changes from Prior IP
Sponsored "hands on" training of biomedical researchers in the use of the computational biology tools at NSF supercomputer center. Topics for workshops on high performance computing included: Connection Machine Parallel Computer, Nucleic Acid and Protein Sequence Analysis, Nuclear Magnetic Resonance.	Topics for scheduled training workshops include: Nucleic Acid and Protein Sequence, Computational X-Ray Crystallography, Heterogeneous Supercomputing Techniques, and Biomechanics Research	Conference on Molecular Modeling and Protein Folding	

NIH	Budget Account Code	Budget (BA, $ M)				IIPCC Components by 1995 Pres. Req.				
		FY 93 Actual	FY 94 Pres.	FY 94 Est.	FY 95 Pres.	HPCS	NREN	ASTA	IITA	BRHR
DCRT High Performance Biomedical Computing Program		8.00	8.40	8.40	11.00	4.00	1.00	3.00	3.00	

The goal of the Division of Computer Research and Technology (DCRT) High Performance Biomedical Computing Program is to make available to the NIH staff the benefits of high performance computing and communication systems in their scientific and clinical research efforts. To achieve this goal, DCRT determines which high performance parallel architectures are best for the classes of problems that arise in biomedical computing, develops parallel algorithms and computational techniques for advanced biomedical computing problems, and provides a high performance parallel computer facility that benefits the NIH staff in their scientific computing needs. DCRT is developing methods and algorithms for a number of biomedical applications that can benefit from computational speedup including image processing of electron micrographs, protein and nucleic acid sequence analysis, nuclear magnetic resonance spectroscopy, x-ray crystallography, protein folding prediction, quantum chemical methods, molecular dynamics simulations, human genetic linkage analysis, medical imaging and radiation treatment planning.

Agency Ties	
ARPA	Partner
NSF	
DOE	
NASA	
NIH	
NSA	
NIST	
NOAA	
EPA	
ED	

FY 1993 Accomplishments	FY 1994 Major Milestones	FY 1995 Major Milestones	Milestone Changes from Prior IP
Progress made in applying high performance parallel computing to biomedical applications including the following: Completed the parallel implementation of the orientation refinement methods used in the three-dimensional reconstruction of icosahedral virus images from electron micrographs. Developed a scalable parallel software package for applying ab initio quantum mechanical methods to chemical systems of biological interest using the Hartree-Fock Self Consistent Field approximation.	Deploy and evaluate a massively parallel computer system capable of providing 10-20 GigaFLOPS for biomedical applications. Initiate the implementation of a distributed computing environment for radiation treatment planning containing workstations for image display and a shared high performance parallel computer for three-dimensional dose calculations. Complete parallel methods for multiple sequence alignment of proteins and nucleic acids. Implement a high-speed network connection to one other high performance computer center in the Washington, D.C. metropolitan area.	Deploy and evaluate a massively parallel computer system capable of providing 20-40 GigaFLOPS for biomedical applications. Complete the implementation of a distributed computing environment for radiation treatment planning containing workstations for image display and a shared high performance parallel computer for three-dimensional dose calculations. Develop new parallel computing methods for reconstructing Positron Emission Tomography (PET) images. Demonstrate high performance parallel computing tools that can be used to discover and design pharmaceuticals based on molecular shape matching techniques.	There are no major milestone changes from the prior IP.

NIH

NCI Frederick Biomedical Supercomputing Center

Budget Account Code	FY 93 Actual	FY 94 Pres.	FY 94 Est.	FY 95 Pres.	HPCS	NREN	ASTA	IITA	BRHR
	6.30	6.70	6.70	6.70	0.90	0.40	4.60		0.80

Budget (BA, $ M); HPCC Components by 1995 Pres. Req.

The NCI's Frederick BiomedicalSupercomputing Center(FBSC) purpose is to provide high performance computing dedicated and available to the entire biomedical scientific community to develop basic knowledge for the diagnosis, treatment, understanding and prevention of cancer and other diseases. It employs advanced techniques in a fully integrated environment of workstations, mid-level, supercomputer and massively parallel computers connected by networks. Activities are concentrated in those areas of biomedical research computation that are too demanding to be pursued on conventional or immature computers. Primary concerns are structure determination by xray and magnetic resonance, structure prediction of nucleic acids and proteins, computational biochemistry and problems that arise from modern molecular biology. Genomic sequence analysis, molecular mechanics, ab initio chemistry, linkage analyis, image analysis and mathematical modeling are primary problem areas. High production algorithms are adapted to vector-multiprocessor and massively parallel systems, entirely new algorithms are developed and leading-edge computer science discoveries from the areas of computer vision, robotics, deterministic and non-deterministic, algorithms.

Agency Ties	
ARPA	
NSF	
DOE	
NASA	
NIH	
NSA	partner
NIST	partner
NOAA	
EPA	
ED	

FY 1993 Accomplishments	FY 1994 Major Milestones	FY 1995 Major Milestones	Milestone Changes from Prior IP
Provided advanced computation to more than 1000 biomedical researchers. Determined the molecular structure of the protease and reverse transcriptase ensymes of HIV and their complexes with prototype drugs. Developed a scalable parallel version of a widely used RNA structure prediction algorithm. Obtained predicted structures for the entire genomes of viruses of the HIV, common cold, poliovirus and other families. Developed a novel, scalable parallel RNA structure prediction program using genetic algorithms.	Improve the productive use of conventional vector-scalar supercomputing techniques through the development of optimized versions of heavily used biomedical applications. Expand the introduction of parallel and distributed processing by continuing to explore use of open tools such as PVM. Apply high performance, parallel computing to diverse, challenging problems of biomedical importance, such as drug development for AIDS and other virus diseases, and cancer.	Use computational and structure-based techniques to design novel inhibitors of enzymes implicated in AIDS, cancer and tuberculosis. Adapt algorithms for motif detection in genomes and molecular structures to several parallel computer architectures. Implement client server technology to improve the system- investigator interfaces.	There are no major milestone changes.

NIH	Budget Account Code	Budget (BA, $ M)				IIPCC Components by 1995 Pres. Req.				
		FY 93 Actual	FY 94 Pres.	FY 94 Est.	FY 95 Pres.	HPCS	NREN	ASTA	IITA	BRHR
NCI High speed networking and distributed conferencing					2.00		0.50	0.40	1.00	0.10

Improve access methodologies for diverse members of the biomedical research community to the entire computational infrastructure of the NCI's Frederick BiomedicalSupercomputing Center(FBSC) by improving the datacommunications and networking infrastructure through the implementation of evolving networking technologies and high speed interfaces. Initiatives in this area cover a wide variety of data communications technologies including local area networking, wide area networking, requirements for multimedia data transmission, dedicated specialized high speed interfaces for local computer to computer connections (e.g. high speedcrossbar switches, HIPPI, fiber channel, etc.), and the use of evolving data communications standards such as ATM and SONET.

Agency Ties	
ARPA	
NSF	
DOE	
NASA	
NIH	
NSA	
NIST	
NOAA	
EPA	
ED	

FY 1993 Accomplishments	FY 1994 Major Milestones	FY 1995 Major Milestones	Milestone Changes from Prior IP
Preliminary planning based on opportunistic access to evolving projects. Developed a network conferencing system running on workstations, connected to the internet for online conferencing to observe, analyze and discuss images of a specific type (gel electropherograms). Developed a scientists workbench for distributing a suite of programs over workstations, supercomputers and mainframes and graphically displaying results.	Evaluate published and unpublished reports on use of high performance networking for local high speed interconnection. Test original in-house methodology, public domain software and commercial products for client/server interfaces to image analysis and conferencing. Seek collaborators in remote sites for visual and computational analysis of biomedical research images.	Acquire and deploy high speed interconnects for the major components of the FBSC. Establish connections to other laboratories of the NIH and biomedical community for high speed data communication. Refine and develop software for access to experimental and computational data from world-wide sources.	

NIH	Budget Account Code	Budget (BA, $ M)				HPCC Components by 1995 Pres. Req.				
		FY 93 Actual	FY 94 Pres.	FY 94 Est.	FY 95 Pres.	HPCS	NREN	ASTA	IITA	BRHR
High Perf. Communications for PDQ, CancerNet, and Electronic Publishing					0.60				0.60	

CancerFax and CancerNet are free services that allow users to access data from NCI databases. CancerFax allows users to dial into one of ICIC's computers from a fax machine and retrieve a faxed image of any of the Physician Data Query (PDQ) statements on cancer screening, treatment, or supportive care in English or Spanish. CancerNet is a free e-mail service on the Internet, enabling users to obtain free access to the same information in CancerFax. Additional cancer information is being added to CancerFax and CancerNet, and utilization is increasing rapidly. Approximately 30% of CancerNet requests originate outside the U.S.

Agency Ties	
ARPA	
NSF	
DOE	
NASA	
NIH	
NSA	
NIST	
NOAA	
EPA	
ED	

FY 1993 Accomplishments	FY 1994 Major Milestones	FY 1995 Major Milestones	Milestone Changes from Prior IP
CancerFax and CancerNet News and database information enhanced. Over 50,000 CancerFax requests and 30,000 CancerNet requests fulfilled on e-mail; current use of CancerNet via gopher and e- mail is approximately 13,000 requests/mon. Access to CancerNet data expanded by implementing on Internet Gopher Service in US, Japan, and Singapore. Electronic mechanism developed for distributing monthly updates of highly focused search output derived from CANCERLIT database on CancerFax and CancerNet. Development begun of totally digital, "platform-independent" multimedia products featuring inter- active info. CDI on cervical cancer produced for distribution and scripts for CDI on breast cancer screening and treatment options partially done.	ICIC will make entire PDQ database accessible over Internet as part of its new Information Associates Program. Prototype for delivering interactive multi-media products will be developed using the Mosaic software. Distribution of cervical cancer CDI through Information Associates Program.	Begin to develop a series of information products that contain text, graphics, sound, and full motion video over the Internet using Mosiac. This will permit NCI to distribute: the Journal of the National Cancer Institute; the Journal of the National Cancer Institute Monographs; OCC educational booklets, and other special reports from the NCI using full multi-media capabilities. Script peer review and programming for CDI on breast cancer screening and treatment options will be completed and CDI prepared for distribution.	

NIH	Budget Account Code	Budget (BA, $ M)				HPCC Components by 1995 Pres. Req				
		FY 93 Actual	FY 94 Pres.	FY 94 Est.	FY 95 Pres.	HPCS	NREN	ASTA	IITA	BRHR
National Coordination Office		0.40	0.90	0.70	0.70					0.70

This program activityis for some of the operation of the National Coordination Office (NCO) for High Performance Computing and Communications. Additional activities are funded by NLM and by participating agencies (for example, agency staff detailed to the NCO).

Agency Ties	
ARPA	
NSF	
DOE	
NASA	
NIH	
NSA	
NIST	
NOAA	
EPA	
ED	

FY 1993 Accomplishments	FY 1994 Major Milestones	FY 1995 Major Milestones	Milestone Changes from Prior IP

National Security Agency

NSA has traditionally influenced and been a very early and sophisticated user of the highest performance commercial computer, storage, and networking systems. For these reasons, NSA actively participated in the original HPCC studies which led to the Federal HPCC program. Through the entire period of growth of high performance computing and networking, spanning several decades, NSA has stimulated both industry and academia with some of the most challenging problems in the nation. A number of major U.S. computer companies are now using hardware and software technology in their products which originated at NSA. This role must continue, both to assure the availability of increasingly higher performance systems to meet the nation's national security interests and to ensure that benefits of NSA's activities accrue to the overall advantage of the industry and the satisfaction of other HPCC Grand Challenges.

NSA will continue to pursue high performance computing and very high speed networks in order to perform its mission. Many of these programs will also contribute directly to the overall goals of HPCC. NSA sponsors the Supercomputing Research Center (SRC), a division of the Institute for Defense Analyses (an FFRDC), to do most of this research.

Program Activity	Budget Account Code	Partner/User Agencies	Budget (BA, $ M)				HPCC Components by 1995 Pres. Req.				
			FY 93 Actual	FY 94 Pres.	FY 94 Est.	FY 95 Pres.	HPCS	NREN	ASTA	IITA	BRHR
Supercomputing Research		ARPA, NASA, NIH		28.30	26.80	26.10	14.10		11.80		0.20
Superconducting Research		NASA		2.00	2.00	2.00	2.00				
Very High Speed Networking				3.50	3.00	3.50		3.50			
Secure Operating System Development		NIST		5.50	5.50	5.70		5.70			
High Speed Data Protection Electronics		ARPA, NASA		2.70	2.70	2.60		2.60			
Technology Based Training				0.23	0.23	0.23				0.23	
		Totals:		42.23	40.23	40.13	16.10	11.80	11.80	0.23	0.20

NSA	Budget Account Code	Budget (BA, $ M)				HPCC Components by 1995 Pres. Req.				
		FY 93 Actual	FY 94 Pres.	FY 94 Est.	FY 95 Pres.	HPCS	NREN	ASTA	IITA	BRHR
Supercomputing Research			28.30	26.80	26.10	14.10		11.80		0.20

The National Security Agency has a perennial requirement for the highest performance computing technology in order to perform its mission, the national security "Grand Challenge." The Supercomputing Research Program is directed to the discovery and application of methods seeking order of magnitude improvements in the computer support required for deriving intelligence from mathematical and signal processing problems. Activities range from invention and prototyping of new concepts, to improvement in the ability to use leading edge commercial products. A current focus of the research is on parallelism, the ability to apply many processing elements simultaneously to a single problem.

The following projects are included within the Supercomputing Research Program:

- PIM - Processor in Memory
- CARNOT - A project to develop heterogeneous data repositories
- SPLASH - A field programmable gate array based architecture
- Distributed Processing - A "farm" of workstation-class systems with very high speed interfaces and switches
- Super Massively Parallel Processor - A "next generation" machine, which combines general purpose vector parallel architecture of the CRAY-3 with two million single-bit PIMs, creating a very high performance massively parallel/vector parallel system.

Agency Ties	
ARPA	Partner
NSF	
DOE	
NASA	Partner
NIH	User
NSA	
NIST	
NOAA	
EPA	
ED	

FY 1993 Accomplishments	FY 1994 Major Milestones	FY 1995 Major Milestones	Milestone Changes from Prior IP
NSA did not appear in the HPCC FY 1993 Implementation Plan.	Develop the PIM concept to a prototype demonstration level. Design large scale PIM-based architecture. Define two CARNOT prototype applications Complete SPLASH development and demonstrate operational proto- types. Transfer SPLASH technology to industry for commercialization. Initiate development of two- dimensional spatial light modulator. Initiate development of two- dimensional photosensitive smart detector. Complete development of C-based compilers. Design a Distributed Processing architecture. Design specific high perfor- mance massively parallel devices. Acquire Performance Analysis prototype system.	Demo Distributed Processing concept. Produce a prototype of a large- scale processor-in-memory-based system integrated into a commercial high perf. vector computer. Explore the use of massively parallel systems on large knowledge- based management applications with inferencing capabilities. Evaluate early parallel process- ing systems with other HPCC agencies and industry. Demonstrate spatial light modulator and photosensitive smart detector component capabilities. Identify and develop additional prototype CARNOT applications. Develop tools and standards to support the open systems concept. Conduct studies of strategies to manage extremely large data files and very high transfer data rates among many different systems.	

84-115 O - 94 - 9

NSA	Budget Account Code	Budget (BA, $ M)				HPCC Components by 1995 Pres. Req.				
		FY 93 Actual	FY 94 Pres.	FY 94 Est.	FY 95 Pres.	HPCS	NREN	ASTA	IITA	BRHR
Superconducting Research			2.00	2.00	2.00	2.00				

The program in superconducting electronics is aimed at providing high performance computing alternatives to current silicon and gallium arsenide technologies, which have speed and power limitations. Prior research suggests that superconducting supercomputers can deliver very high performance with very low power requirements. NSA would like to cooperate with industry to develop such a computer.

Agency Ties	
ARPA	
NSF	
DOE	
NASA	Partner
NIH	
NSA	
NIST	
NOAA	
EPA	
ED	

FY 1993 Accomplishments	FY 1994 Major Milestones	FY 1995 Major Milestones	Milestone Changes from Prior IP
NSA did not appear in the HPCC FY 1993 Implementation Plan.	Develop a prototype superconducting crossbar switch.	Install and test the prototype superconducting crossbar switch in a working system. Develop superconducting electronics technology which could be applied to commercial supercomputers of the future. Study feasibility of building a general purpose computer using superconducting technology.	

NSA	Budget Account Code	Budget (BA, $ M)				HPCC Components by 1995 Pres. Req.				
		FY 93 Actual	FY 94 Pres.	FY 94 Est.	FY 95 Pres.	HPCS	NREN	ASTA	IITA	BRHR
Very High Speed Networking			3.50	3.00	3.50		3.50			

The National Security Agency has a perennial requirement for the fastest networking technology in order to perform its mission, the national security "Grand Challenge." The Very High Speed Networking Program will provide NSA a high performance network infrastructure characterized by both multi-gigabit per second trunking speeds and the ability to support sustained data flows of at least hundreds of megabits per second each now and ultimately multi-gigabits per second each.

Agency Ties	
ARPA	
NSF	
DOE	
NASA	
NIH	
NSA	
NIST	
NOAA	
EPA	
ED	

FY 1993 Accomplishments	FY 1994 Major Milestones	FY 1995 Major Milestones	Milestone Changes from Prior IP
NSA did not appear in the HPCC FY 1993 Implementation Plan.	Install an in-house Gb/s network testbed to explore high-speed network architectures and network management techniques. There will be a similar testbed capability to explore compatibility of DoD private networks with commercial public networks. Collaborate with industry on the development of high-speed (electronic and optical) prototype switching equipment. Collaborate with industry on the development of operation, administration and management protocols for optically switched networks.	Demonstrate electronic switching of up to 2.4 Gb/s data streams. Initiate development of optical switching and wave division multiplexing devices to support 2.4 Gb/s data rates. Conduct soliton research to explore application to high-speed networks.	

NSA

Secure Operating System Development

Budget Account Code	FY 93 Actual	FY 94 Pres.	FY 94 Est.	FY 95 Pres.	HPCS	NREN	ASTA	IITA	BRHR
		5.50	5.50	5.70		5.70			

Budget (BA, $ M) — FY 93 through FY 95; HPCC Components by 1995 Pres. Req. — HPCS, NREN, ASTA, IITA, BRHR.

NSA is responsible for researching solutions to ensure the security of the nation's networks and distributed systems. In collaboration with NIST and DISA, NSA is identifying issues and researching security policy-flexible, cost-effective ways to ensure the security of both government and commercial enterprise systems. The Synergy research program at NSA is developing an "open architecture" along with secure distributed system prototypes based upon security policy-flexible, operating system microkernels. Synergy will provide a means for commercial vendors to address a wide variety of security markets with a single architecture, thus lowering everyone's costs. Synergy can be the foundation on which security solutions for major initiatives can be built, whether addressing government-peculiar needs or commercial-enterprise needs, from the Defense Department Global Grid to the National Information Infrastructure. Synergy will integrate the INFOSEC research work in computer misuse and anomaly detection (audit/intrusion detection), real-time and multimedia, availability, network security management, high speed networking, and secure database management systems.

Agency Ties	
ARPA	
NSF	
DOE	
NASA	
NIH	
NSA	
NIST	Partner
NOAA	
EPA	
ED	

FY 1993 Accomplishments	FY 1994 Major Milestones	FY 1995 Major Milestones	Milestone Changes from Prior IP
NSA did not appear in the HPCC FY 1993 Implementation Plan.	Establish contracts or CRADAs with major commercial vendors and begin to implement Synergy architecture on their microkernels to examine portability, performance, security policy flexibility issues. Prototype Synergy security policy flexible file and network servers. Begin experiment to run Synergy-based architecture over ATM high-speed network. Begin integration of network management concepts to manage Synergy-based security components. Initiate 5-10 research contracts with universities to advance Synergy research and generate new ideas. Demo Synergy prototype architecture's flexibility and applicability to commercial or NII applications.	Release Synergy proof of concept software, to universities, commercial research labs, and secure system vendors. Continue university research program to advance Synergy arch. Fund 1- 3 universities to develop course in secure distributed systems and perform Synergy research. Initiate internal and external research efforts to mitigate the risks and costs associated with high assurance product implementations of the Synergy architecture. Develop a centralized security management tool to be integrated with Synergy on a high-speed testbed. Establish a collaborative relationship via CRADA or contract with one or more commercial operating system vendors or consortia, to begin implementing the Synergy architecture in a commercial distributed system.	

NSA	Budget Account Code	Budget (BA, $ M)				HPCC Components by 1995 Pres. Req.				
		FY 93 Actual	FY 94 Pres.	FY 94 Est.	FY 95 Pres.	HPCS	NREN	ASTA	IITA	BRHR
High Speed Data Protection Electronics			2.70	2.70	2.60		2.60			

Network data security is a requirement throughout the National Community, from the financial community through the medical community to a wide range of research activities. NSA will take the lead in developing network security and information security techniques and products for high speed networks. It will establish high speed network testbeds to explore network security issues. Many of the research efforts have moved from protection of network trunks to protection of high-speed individualized computer links. The focus of this effort is to develop a technology which will interface to network management systems that can be used with the high-speed networks that are exemplified by NREN.

Agency Ties	
ARPA	Partner
NSF	
DOE	
NASA	Partner
NIH	
NSA	
NIST	
NOAA	
EPA	
ED	

FY 1993 Accomplishments	FY 1994 Major Milestones	FY 1995 Major Milestones	Milestone Changes from Prior IP
NSA did not appear in the HPCC FY 1993 Implementation Plan.	Design, fabricate, and test ICs with complexities ranging from simple test structures to large-scale integration (1,000 to 100,000 gates per chip). Develop the basic system design needed to secure networks at 622 Mb/s with extensibility to 2.4 Gb/s and develop test structures in areas determined to be high risk. Continue research in bulk and integrated optical technologies. Develop small arrays of smart pixels. Develop the optical interconnection system needed to route optical signals between smart pixel array planes.	Extend the density and IC complexity to very large-scale integration levels of 100,000 to 1,000,000 gates. Develop the detailed design for secure communication applied to ATM (622 Mb/s system) including the design of all custom ICs needed. The first fabrication and testing of the system ICs will also take place. Begin development of a prototype 2.1 Gb/s front end processor for a high speed data encryption unit. Develop and demonstrate a 3.2 Gb/s optical interconnection system.	

NSA	Budget Account Code	Budget (BA, $ M)				HPCC Components by 1995 Pres. Req.				
Technology Based Training		FY 93 Actual	FY 94 Pres.	FY 94 Est.	FY 95 Pres.	HPCS	NREN	ASTA	IITA	BRHR
			0.23	0.23	0.23				0.23	

NSA considers lifelong access to education and training for all employees to be highly critical in order to maintain and enhance their skills and ensure our viability in a world of rapidly changing technology. Traditional training approaches take too long and are too expensive. Consequently, we are developing a skill-based assessment and training model that will use emerging communications and computer techniques to sustain each employee's abilities at the levels required in today's constantly changing workplace. The key ingredient in this approach will be on-demand access to Technology Based Training (TBT) of known quality and relevance delivered directly to the workplace. Delivery over corporate LANs and the INTERNET, and the possible use of ISDN technology will be addressed. While initial offerings would consist largely of text-based CBT courseware, digital video and other media will be accommodated as they become available.

Placing existing TBT courseware on the network is something that can be accomplished quickly, but there are various conceptual, technical and procedural issues that must be addressed before it can be made generally accessible. Foremost among these are:

1- Define the skills needed to perform specific jobs and a process for certifying the efficacy of training modules that address those skills.
2- Storage, access, and maintenance of large numbers of TBT courses.
3- Cataloging of available courseware in a manner that minimizes the difficulty of identifying suitable courses or training modules.
4- Identifying and resolving the interoperability problems in delivering TBT courses over a network to a wide variety of computer terminals.
5- Defining and selecting appropriate user interfaces. Special effort will be focused on interfaces for home-computer configurations.
6- Developing an EDI based business model bringing market forces to bear on the availability and costs of Technology Based Training.

Agency Ties	
ARPA	
NSF	
DOE	
NASA	
NIH	
NSA	
NIST	
NOAA	
EPA	
ED	

FY 1993 Accomplishments	FY 1994 Major Milestones	FY 1995 Major Milestones	Milestone Changes from Prior IP
NSA did not appear in the HPCC FY 1993 Implementation Plan.	Implement pilot demonstration of TBT over the NSA corporate network. Demonstrate delivery of TBT over INTERNET. Research/select skill-based job performance model. Assemble executive level government-industry group to establish certification processes for skill-based training modules.	Perform skill-based certification of ca. 100 existing training modules. Build user-friendly catalog of available certified courses. Establish "digital warehouses" of TBT on-line via INTERNET and the NSA corporate network. Implement pilot of EDI-based business model.	This entire activity is a new effort within the NSA HPCC Program.

National Institute of Standards and Technology

Goals of NIST's HPCC Program are:

* Apply high performance communications and networking technology to promote improved U.S. product quality and manufacturing performance, to reduce production costs and time-to-market, and to increase competitiveness in international markets;
* Promote the development and deployment of advanced communications technology to support the education, research and manufacturing communities and to increase the availability of scientific and engineering data via the NII;
* Advance instrumentation and performance measurement methodologies for high performance computing and networking systems and components, and to achieve improved system and application performance;
* Develop efficient algorithms and portable, scalable software for the application of high performance computing systems to industrial problems, and to develop improved methods for the public dissemination of advanced software and documentation; and
* Support, promote, and coordinate the development of voluntary standards that provide interoperability and common user interfaces among systems.

Objectives of NIST's HPCC Program are to accelerate the development and deployment of high performance computing and networking technologies required for the National Information Infrastructure, and to apply and test these technologies in a manufacturing environment.

HPCS -- The HPCS program component focusses on instrumentation and methodology for performance measurement of scaleable, high performance computing and networking systems, and on meeting metrology needs of future generations of microelectronics. Emphasis of the performance measurement component is on the development and use of low perturbation data capture hardware and simplified software-based approaches to performance characterization of these systems and software components. The microelectronics component supports R&D leading to measurement methods and systems which will be employed by industry to manufacture products.

NREN -- NIST supports and coordinates the development of standards within the Federal government to provide interoperability, common user interfaces to systems, and enhanced security. NIST works with other agencies to promote open system standards and to aid in the commercialization of technology by U.S. industry. NIST promotes the development of communications infrastructure and through information technology research, development, and related activities to enhance basic communications capabilities. Supporting activities include metrology to assess network components for performance, interoperability and conformity; metrology to enable the manufacture of component technologies; and collaboration with industry via cooperative research and development, sponsorship of workshops, other public fora, and laboratory-based technology demonstrations.

ASTA -- NIST develops algorithms and generic software for advanced scientific, engineering and manufacturing applications. Common elements and techniques are encapsulated in software libraries to promote ease of use and application portability. NIST's Guide to Available Mathematical Software (GAMS) provides industry and the public with improved electronic access to reusable software.

IITA -- NIST will build upon its experience in information technology and manufacturing engineering to accelerate the application of high performance computing and communication technology to manufacturing environments. NIST will support expanded programs in: advanced manufacturing systems integration technologies; development and test of prototype components and interface specifications for manufacturing systems; application of high performance computing and networking technologies to integrate the design and production processes; and testbeds for achieving cost-effective application of advanced manufacturing systems and networks. NIST will also address key issues in electronic commerce, electronic libraries, and network security.

National Institute of Standards and Technology

Program Activity	Budget Account Code	Partner/User Agencies	Budget (BA, $ M)				HPCC Components by 1995 Pres. Req.				
			FY 93 Actual	FY 94 Pres.	FY 94 Est.	FY 95 Pres.	HPCS	NREN	ASTA	IITA	BRHR
NREN Deployment and Performance Measures for Gigabit Nets & MPP Sys.	STRS for Computer Sys. 370	ARPA, DOE, NOAA, EPA	1.50	3.00	2.20	2.20	0.30	1.90			
Development and Dissemination of Scientific Software for HPCS	STRS for Computer Sys. 370	NSF, DOE, NASA, EPA	0.60	4.10	3.60	7.60			4.60	3.00	
Support for Electronic Commerce	STRS for Computer Sys. 370			2.50	0.95	2.75				2.75	
Electronic Libraries and Distributed Multimedia Applications	STRS for Computer Sys. 370			2.00	0.85	1.25				1.25	
Systems Integration for Manufacturing Applications	STRS for Computer Sys. 370	ARPA, DOE		12.80	9.00	25.20				25.20	
Assurance, Reliability and Integrity of NREN Objects	STRS for Computer Sys. 370	NSA				1.20				1.20	
Specification and Testing of High Integrity, Distributed Systems	STRS for Computer Sys. 370			1.70		4.00				4.00	
Language, image and text processing	STRS for Computer Sys. 370	ARPA, NSA			1.50	4.00				4.00	
Metrology for Future Generations of Microelectronics	STRS for Computer Sys. 370	DOE				6.45	6.45				
Metrology to support mobile and fixed based communications networks	STRS for Computer Sys. 370					1.75		1.75			
		Totals:	2.10	26.10	18.10	56.40	6.75	3.65	4.60	41.40	

NIST	Budget Account Code	Budget (BA, $ M)				HPCC Components by 1995 Pres. Req.				
		FY 93 Actual	FY 94 Pres.	FY 94 Est.	FY 95 Pres.	HPCS	NREN	ASTA	IITA	BRHR
NREN Deployment and Performance Measures for Gigabit Nets & MPP Sys.		1.50	3.00	2.20	2.20	0.30	1.90			

Promote the development of communications infrastructure and its use via information technology research, development, and related activities to enhance basic communications capabilities. Support and coordinate the development of communications standards within the Federal government to provide interoperability, common user interfaces to systems, and enhanced security; work with other agencies to promote open system standards to aid in the commercialization of technology by U.S. industry. NIST will utilize its expertise to enable transition planning and deployment of ISDN and OSI-based protocols in the NREN and undertake R&D to support emerging Broadband ISDN standards.

NIST developed the MultiKron instrumentation chip to achieve low-perturbation data capture while assessing the performance of scalable high performance computer systems and high speed networking technologies. Performance evaluation activities include measurement and characterization of the impact of software protocols on communications performance in order to minimize communications bottlenecks between application programs. NIST's researcher have implemented a high speed communications testbed that integrates supercomputers, HIPPI interfaces, workstations, and SONET communications technologies, and employs NIST's VLSI-based MultiKron performance measurement chip to investigate the behavior and performance of communications protocols suitable for multi-gigabit/sec. transmission rates. NIST collaborates with industry, academia and other interested agencies.

Agency Ties	
ARPA	Partner
NSF	
DOE	Partner
NASA	
NIH	
NSA	
NIST	
NOAA	User
EPA	User
ED	

FY 1993 Accomplishments	FY 1994 Major Milestones	FY 1995 Major Milestones	Milestone Changes from Prior IP
Sponsorship of North American ISDN Users Forum continues; industry and users met four times to address implementation issues. NIST designed, developed and has shared a designer and experimenter's kit which allows the Multikron instrumentation to be integrated into systems using the industry standard VME bus. Devised a software-based technique for portable, sensitivity-to performance measurements of software executing on scalable parallel computing systems. Applied Multikron instrumented systems to assess performance of TCP/IP Protocols over a HIPPI-based network (High Performance Parallel Interface) and published results.	Apply Mulitkron-based performance measurement technology to Asynchronous Transfer Mode (ATM) based networks to identify bottlenecks and publish results. Conduct simulation studies of ATM-based protocols and networks to ascertain dynamic behaviors and performance; contribute results to appropriate standards bodies including the ATM Forum, Am. Nat. Standards Telecommunications Committee (T1S1), and the International Telecommunications Union, Telecom Sector Committee. Continue to sponsor the ISDN User's Forum; facilitate the ISDN demonstration entitled: "ISDN Solutions -94-" and participate in it.	Collaborate with the ATM Forum to develop standardized conformity tests for ATM-based technology. Collaborate with industry and users to establish an ATM testbed; conduct experiments and tests of evolving products. Conduct R&D related to media access and flow control for ATM protocols; and congestion control algorithm. Publish results and share in appropriate standards fora.	N/A

NIST	Budget Account Code	Budget (BA, $ M)				HPCC Components by 1995 Pres. Req.				
		FY 93 Actual	FY 94 Pres.	FY 94 Est.	FY 95 Pres.	HPCS	NREN	ASTA	IITA	BRHR
Development and Dissemination of Scientific Software for HPCS		0.60	4.10	3.60	7.60			4.60	3.00	

Develop advanced mathematical, computational and visualization algorithms, software, methodology and tools which support the efficient application of computationally intensive science to key problems arising in the industrial sector. Current application areas addressed include the development of improved computational methods for chemical processing and related advanced materials processing industries. Topics of emphasis are methods for improved advanced product and process design and the generation of computationally derived fundamental data for environmentally sound competitive manufacturing processes. Technology areas of emphasis include the interactive visualization of complex structures and data, parallel algorithms, large-scale Ab Initio and semi-empirical computational chemistry, Monte Carlo and molecular dynamics modeling and scientific database utilization.

This program activity also supports the development of modern, network-based reusable software classification and distribution technology for making new computational software readily available to industry and the public. The focus of this activity is the NIST Guide to Available Mathematical Software (GAMS) project. GAMS is a working virtual software repository which provides users convenient access to thousands of software modules which currently reside in four separate repositories.

Agency Ties	
ARPA	
NSF	Partner
DOE	Partner
NASA	Partner
NIH	
NSA	
NIST	
NOAA	
EPA	User
ED	

FY 1993 Accomplishments	FY 1994 Major Milestones	FY 1995 Major Milestones	Milestone Changes from Prior IP
Electronic access by the public and industry to NIST's Guide to Available Mathematical Software (GAMS) reusable software database was augmented with expanded network access, and an improved graphical user interface. GAMS software coverage was extended to over 8800 software modules in 70 mathematical and statistical software packages, including 200 additional modules from the netlib repository of Oak Ridge National Laboratory. Improved methods for evaluating and characterizing parallel random number generators were developed based upon a mapping of the point set to one dimensional strings.	Develop and implement initial parallel algorithms for applications in computational chemistry including Monte Carlo methods, molecular dynamics, fluid dynamics, interactive molecular graphics and quantum chemistry. Develop scalable algorithms for the computation of spin-dependent electron transport properties of magnetic multilayer structures used in magnetic storage devices. Develop prototype visualization tools and scalable algorithms which support common needs in computational chemistry and materials properties. Enhance the GAMS system interface and develop a client/server architecture to provide a software dissemination testbed.	Develop and implement methods for applying quantum chemistry to the prediction of properties and reactivities of solvated systems, including the introduction of discrete and continuum solvent representations into modern quantum chemistry codes and the use of molecular dynamics and Monte Carlo methods to explore the energetics of solvent/solute clusters. Develop scalable quantum molecular dynamics software for modeling growth of metal films on semiconductor surfaces, and apply to optimizing the performance of x-ray multilayer mirrors. Demonstrate a useful and reliable GAMS service for the access and retrieval of reusable software which supports automated expert advice on high performance computer algorithms and software.	N/A

NIST	Budget Account Code	Budget (BA, $ M)				IIPCC Components by 1995 Pres. Req.				
		FY 93 Actual	FY 94 Pres.	FY 94 Est.	FY 95 Pres.	HPCS	NREN	ASTA	IITA	BRHR
Support for Electronic Commerce			2.50	0.95	2.75				2.75	

Collaborate with industry and other agencies to develop or apply technologies which enable electronic commerce in general, and as it applies to manufacturing electronic and mechanical components and subsystems, including electronic information query, dissemination and subsequent integration of product specifications into computer aided design and manufacturing tools. Establish supporting laboratory facilities to support demonstrations, evaluation, and inter-operation testing. Conduct R&D to support development of tools which generate software for electronic commerce (EC) applications; develop an EC integration facility at NIST; develop requirements and prototype systems for handling complex data in Electronic Data Interchange (EDI); and design and develop prototype interfaces among EDI, Structured Query Language (SQL), and Remote Database Access (RDA). Also, conduct R&D leading to development of application programming interfaces for digital signature, authentication, and other security services required to support electronic commerce. Establish cooperative R&D agreements with researchers, vendors and users to better understand requirements and to acquire, test, interconnect and demonstrate prototype systems. Collaborations will include DOE: Lawrence Livermore and Sandia National Labs; DOD/Defense Logistics Agency; Dept. of Treasury; and other interested agencies.

Agency Ties	
ARPA	
NSF	
DOE	
NASA	
NIH	
NSA	
NIST	
NOAA	
EPA	
ED	

FY 1993 Accomplishments	FY 1994 Major Milestones	FY 1995 Major Milestones	Milestone Changes from Prior IP
N/A	Develop an Electronic Commerce integration and demonstration facility at NIST; solicit industry participation and demonstration of products. Prepare for and complete basic functionality of electronic procurement pilot project. Determine interoperability issues related to use of value added networks and electronic mail. Determine and initiate development or integration of security services needs e.g., digital signature, authentication services. Develop prototype software which can generate interfaces and mapping software for EDI applications given a description of a transaction set; demonstrate selected applications.	Demonstrate electronic procurement and publish application guidelines for government. Continue development and integration of required security services. Provide specifications for on-line catalogs for electronic and mechanical components. Extend the demonstration facilities to support ad hoc query and response to trading partner databases via EDI/SQL/RDA-based standards. With industry, publish a draft report describing a strategy to use the Standard for the Exchange of Product Model Data (STEP), Initial Graphics Exchange Specification (IGES), Computer Graphics Metafile (CGM) and Programmers Hierarchical Interactive Graphics System (PHIGS) for management and presentation of product data.	N/A

NIST

Electronic Libraries and Distributed Multimedia Applications

Budget Account Code	FY 93 Actual	FY 94 Pres.	FY 94 Est.	FY 95 Pres.	HPCS	NREN	ASTA	IITA	BRHR
		2.00	0.85	1.25				1.25	

Establish an electronic library pilot project which includes the creation, editorial review, production and dissemination of scientific and engineering papers, documents, data and other published information in a working, production environment via the library and scientific and engineering operating units at NIST. Within this environment provide for the organization, search and retrieval, and dissemination of information by electronic means using state-of-the-art definition and dissemination technologies for documents and multimedia objects, particularly device-independent, portable, exchange formats for information objects.

Within this context, determine the adequacy of individual multimedia and database standards and possible extensions based on the application of object technology to multimedia; develop a profile or architectural framework describing the required standards and their interrelationships; develop standards, prototypes, and demonstrations in collaboration with industry, academia and other interested parties through cooperative research and development agreements.

Collaborate with other parties within academia, the library and information communities, other agencies and industry to investigate the framework for an electronic, distributed, national engineering library.

Agency Ties	
ARPA	
NSF	
DOE	
NASA	
NIH	
NSA	
NIST	
NOAA	
EPA	
ED	

FY 1993 Accomplishments	FY 1994 Major Milestones	FY 1995 Major Milestones	Milestone Changes from Prior IP
N/A	Establish a pilot project at NIST to evaluate alternates to traditional paper-based authoring, publishing and information dissemination. Establish a cooperative R&D agreement with Adobe to assess Acrobat/ Portable Document Format as a document interchange format. Initiate R&D on the communications services needed to support transfer of multimedia information objects. Establish initial collaborations with other interested parties.	Integrate and demonstrate the use of Portable Document Format (PDF) via network-based support tools like Mosaic. Establish a formal integration process to enable authors to create and disseminate PDF-based documents via Mosaic accessible information bases. Develop and publish plans to establish a model for a national scientific, technical and engineering information resource; draft initial specifications for infrastructure and information content.	N/A

NIST	Budget Account Code	Budget (BA, $ M)				HPCC Components by 1995 Pres. Req.				
		FY 93 Actual	FY 94 Pres.	FY 94 Est.	FY 95 Pres.	HPCS	NREN	ASTA	IITA	BRHR
Systems Integration for Manufacturing Applications			12.80	9.00	25.20				25.20	

Emphasis is on systems integration technologies that support flexible and rapid access to information for manufacturing applications. A standards-based data exchange effort for computer integrated manufacturing focuses on improving data exchange among design, planning and production activities. Prototype systems and interface specifications will be communicated to standards organizations. Results will be made available to US industry through workshops, training materials, electronic data repositories and pre-commercial prototype systems that can be installed by potential vendors for test and evaluation in NIST and industry supported Advanced Manufacturing Systems and Networking Testbeds. NIST will distribute standards reference data, technical information, and digital product data designs via digital library technologies.

NIST's manufacturing testbed will support R&D in high performance manufacturing systems and testing computer and networking hardware and software in a manufacturing environment, and will be extended to include applications in mechanical, electronics, construction, and chemical industries; electronic commerce applications in mechanical and electronic products; apparel manufacturing; integration of materials performance data into computer aided design; harmonization of product data exchange and electronic data interchange standards; technology transfer of information for electronics and construction industries; and an integrated Standard Reference Data system. The testbed will serve as a demonstration site for use by industrial technology suppliers and users, and assist industry in the development and implementation of voluntary standards.

The work will support cooperative research and development agreements with industry and academic researchers.

Agency Ties	
ARPA	Partner
NSF	
DOE	Partner
NASA	
NIH	
NSA	
NIST	
NOAA	
EPA	
ED	

FY 1993 Accomplishments	FY 1994 Major Milestones	FY 1995 Major Milestones	Milestone Changes from Prior IP
N/A	Install and evaluate state-of-the-art manufacturing systems for design and planning of mechanical part production. Implement a prototype application protocol development environment for product data exchange standards. Publish report on reference architecture for manufacturing control systems and interface specifications for production systems. Provide electronic technology transfer capabilities for NIST's manufacturing projects in response to industry request for online research information for advanced manufacturing systems.	Transfer interface specifications for production control systems into industry developed testbeds for verification purposes. Evaluate additional state-of-the-art software systems for scheduling and production control of mechanical parts production. Publish requirements for a computer-aided manufacturing engineering tools environment. Specify application protocols for apparel and electronics mfg. Publish a reference model for process plant design Implement a manufacturing testbed with the apparel industry. Demonstrate a minimum set of conformance testing capabilities for the Standard for the Exchange of Product Data (STEP). Make selected standard reference data available via an improved human-machine interface.	N/A

NIST	Budget Account Code	Budget (BA, $ M)				HPCC Components by 1995 Pres. Req.				
		FY 93 Actual	FY 94 Pres.	FY 94 Est.	FY 95 Pres.	HPCS	NREN	ASTA	IITA	BRHR
Assurance, Reliability and Integrity of NREN Objects					1.20				1.20	

Conduct R&D regarding the applicability of object oriented design and development methodology to development of security technology. Extend and apply the Common Criteria which is being developed by governments in North America and Europe. Develop and publish criteria for specifying security requirements, methods for assuring requirements are met, and methods to perform certification and accreditation; and initiate development of an information technology security accreditation program.

Agency Ties	
ARPA	
NSF	
DOE	
NASA	
NIH	
NSA	Partner
NIST	
NOAA	
EPA	
ED	

FY 1993 Accomplishments	FY 1994 Major Milestones	FY 1995 Major Milestones	Milestone Changes from Prior IP
N/A	N/A	Conduct R&D on object oriented development methods for improving security, reliability, and integrity requirements of objects. Publish guidance on the use of Common Criteria for specification of security, reliability and integrity requirements. Create and publish a set of profiles which specify security, reliability, and integrity requirements that are applicable to NREN/NII objects.	N/A

NIST	Budget Account Code	Budget (BA, $ M)				HPCC Components by 1995 Pres. Req.				
		FY 93 Actual	FY 94 Pres.	FY 94 Est.	FY 95 Pres.	HPCS	NREN	ASTA	IITA	BRHR
Specification and Testing of High Integrity, Distributed Systems			1.70		4.00				4.00	

Conduct R&D to extend methods to formally specify, verify and automate testing for conformity and interoperability of high integrity, secure, communications protocols and distributed software systems. Evaluate object oriented methods and technologies for specification of high integrity systems. Develop and extend computer aided design, analysis and testing methodologies to support protocol and software engineering through the use of object oriented technologies. Software engineering environments (SEEs) are specifically designed to facilitate software design, validation, production and maintenance. But, SEEs suffer from problems in interoperability and information sharing where the tools do not readily share information or communicate with each other. Conduct research, develop test-suites, and demonstrate and validate concepts required to support an integrated software engineering environment. Identify required software tool interfaces to promote applications interoperability and portability across a distributed software engineering environment.

Also, conduct R&D to define a process by which an organization: defines, specifies and manages the application of distributed systems to its operation; identifies standards and component technologies which can be used to assemble distributed systems; and establish a testbed to manage federated systems (distributed systems owned by more than one entity).

Agency Ties	
ARPA	
NSF	
DOE	
NASA	
NIH	
NSA	
NIST	
NOAA	
EPA	
ED	

FY 1993 Accomplishments	FY 1994 Major Milestones	FY 1995 Major Milestones	Milestone Changes from Prior IP
N/A	N/A	Conduct workshops with industry, academia and government. Establish laboratory facilities to conduct R&D and assess candidate technologies which support formal specification, automated testing, and integrated software engineering. Collaborate with industry and standards bodies to support the specification, development, validation and testing of standards and conforming products. Assess requirements for integrated software engineering tools and object oriented specification technologies. Publish a framework for object oriented services supporting integrated software engineering environments; solicit review and comment. Conduct R&D to establish a distributed systems definition methodology for management of federated systems.	N/A

NIST	Budget Account Code	Budget (BA, $ M)				HPCC Components by 1995 Pres. Req.				
		FY 93 Actual	FY 94 Pres.	FY 94 Est.	FY 95 Pres.	HPCS	NREN	ASTA	IITA	BRHR
Language, image and text processing				1.50	4.00				4.00	

Conduct cooperative R&D and promote the development of algorithms and related reference materials necessary to conduct research, development, and commercialization of such research for target application areas in speech understanding, image recognition, and intelligent text retrieval. NIST will collaborate with the research communities in spoken natural language, printed and cursive handwriting recognition, and text search and retrieval by providing speech, image and text corpora as reference materials, and developing performance measures and scoring protocols.

Agency Ties	
ARPA	Partner
NSF	
DOE	
NASA	
NIH	
NSA	Partner
NIST	
NOAA	
EPA	
ED	

FY 1993 Accomplishments	FY 1994 Major Milestones	FY 1995 Major Milestones	Milestone Changes from Prior IP
N/A	Implement a prototype wide area information service (WAIS) using NIST's Probabilistic Ranking Information Search Engine (PRISE) system to access the data. The PRISE system is a search engine that has proven user effectiveness on large collections of text. Develop optical character recognition (OCR) techniques to segment bitmapped page images into components (text, figures, tables, etc.). Establish a demonstration library to serve manufacturing interests, incorporating robust user searching methods, and providing access to CALS (Computer Aided Logistics Standards (CALS) information.	Develop an initial dialogue management module which will provide a spoken-language query interface to library information databases. Develop initial algorithms for understanding of documents and associated metrics for judging algorithm performance. Develop a reference implementation of an enhanced wide area information service with appropriate protocol extensions for probabilistic search criteria (WAIS/Z39.50+). Initiate investigations with OCR techniques to segment multimedia information sources.	N/A

NIST	Budget Account Code	Budget (BA, $ M)				HPCC Components by 1995 Pres. Req.				
		FY 93 Actual	FY 94 Pres.	FY 94 Est.	FY 95 Pres.	HPCS	NREN	ASTA	IITA	BRHR
Metrology for Future Generations of Microelectronics					6.45	6.45				

Provide advanced measurement support for semiconductors and microelectronics and the measurement standards that are critical for such technologies. The support for semiconductor and microelectronics focuses on measurement technology to: (1) control fabrication processes in real-time to produce more complex systems; (2) characterize semiconductor materials with greater precision and over the smaller dimensions that are needed for faster systems; and (3) support industry's development of advanced lithography to increase both speed and functionality. These efforts will ensure availability of advanced technologies required for high performance components and systems and will provide a metrological foundation for the more powerful and measurement demanding systems to follow. It also enables U.S. industry to produce the basic approaches for complex configurations of workstations, advanced computers, network servers, and networks.

Agency Ties	
ARPA	
NSF	
DOE	Partner
NASA	
NIH	
NSA	
NIST	
NOAA	
EPA	
ED	

FY 1993 Accomplishments	FY 1994 Major Milestones	FY 1995 Major Milestones	Milestone Changes from Prior IP
N/A	N/A	Submit draft standards for the activation energy of electro- migration and for the acceleration factor for current-density stress to the Joint Electron Device Engineering Council. Demonstrate in-situ process monitoring of film resistivity and thermal history during film deposition. Conduct 2-D laser scatter imaging of silicon etching plasmas and correlate to other plasma measurements as the first comprehensive 2-D measurements ever obtained in a well characterized plasma reactor. Select and test components for wafer-plane spectroradiometers, including array detectors, filters, monochromators, light collectors, diffusers, and fiber optics. Select light sources to test as calibration sources.	N/A

NIST	Budget Account Code	Budget (BA, $ M)				HPCC Components by 1995 Pres. Req.				
		FY 93 Actual	FY 94 Pres.	FY 94 Est.	FY 95 Pres.	HPCS	NREN	ASTA	IITA	BRHR
Metrology to support mobile and fixed based communications networks					1.75		1.75			

Provide advanced measurement support for advanced wired, optical, wireless and mixed telecommunications media. Initial metrology will characterize high-performance microwave integrated circuits and will provide for the timing of very high speed signals required for future mobile and fixed based networks. The support for microwave integrated circuits emphasizes wafer-level standards, miniature individual components, and the dielectric and magnetic materials used in these components. The support for telecommunications timing systems includes enhanced modeling of degradation of synchronization of optical networks, analyzing and measuring phase noise in such systems, coordinating a U.S. industry position on timing distribution, and advocating the latter through international committees.

Agency Ties	
ARPA	
NSF	
DOE	
NASA	
NIH	
NSA	
NIST	
NOAA	
EPA	
ED	

FY 1993 Accomplishments	FY 1994 Major Milestones	FY 1995 Major Milestones	Milestone Changes from Prior IP
N/A	N/A	Enhance impedance characterization methods to allow characterization of crosstalk in multiple, lossy, high-speed, electronic interconnections using both frequency and time domain techniques. Demonstrate utility of methods using commercial components. Submit methods for adoption as industry standards. Develop more efficient and simpler methods and systems for phase-noise characterization of synchronization components and systems proposed for the gigabit-per-second data rates.	N/A

National Oceanic and Atmospheric Agency

NOAA's Grand Challenge research in climate prediction and weather forecasting depends on advances in high-end computing and on the collection and dissemination of environmental information. Increased computing power will enable more accurate representation of the atmosphere-ocean system, resulting in improved weather forecasts and making possible better decision making by Government and industry on issues that affect both the environment and the economy. Increased use of the Internet and follow-on networks of the National Information Infrastructure will enable NOAA to disseminate its vast holdings of real-time and historical information to all users more completely, in a more usable form, and in a much more timely manner.

Program Activity	Budget Account Code	Partner/User Agencies	Budget (BA, $ M)				HPCC Components by 1995 Pres. Req.				
			FY 93 Actual	FY 94 Pres.	FY 94 Est.	FY 95 Pres.	HPCS	NREN	ASTA	IITA	BRHR
Advanced Computation	ORF EI		9.40	10.60	9.85	16.05			16.05		
Networking Connectivity	ORF EI		0.40	1.80	0.95	8.70		8.70			
Information Dissemination Pilots	ORF EI					0.50				0.50	
		Totals:	9.80	12.40	10.80	25.25		8.70	16.05	0.50	

NOAA	Budget Account Code	Budget (BA, $ M)				HPCC Components by 1995 Pres. Req.				
Advanced Computation		FY 93 Actual	FY 94 Pres.	FY 94 Est.	FY 95 Pres.	HPCS	NREN	ASTA	IITA	BRHR
		9.40	10.60	9.85	16.05			16.05		

Advanced Computation is NOAA's program to make possible major improvement in the Nation's ability to forecast the weather and predict climate change by taking full advantage of highly parallel, high performance computing systems which are, over the long term, expected to provide substantially greater computing power at lower cost.

NOAA/GFDL's collaborative efforts with DOE/LANL and the Naval Post Graduate School have resulted in an extremely high resolution global ocean model, built largely on the GFDL Modular Ocean Model (MOM) that shows detail not yet achieved using conventional computing resources and a global atmospheric grid-point model which is important for the study of climate, middle atmosphere dynamics, and atmospheric ozone.

NOAA/NMC's collaborative efforts wsith DOD/NRL have resulted in execution of a parallel, adiabatic version of the NMC global spectral model with excellent performance at high resolution.

The installation of the Intel Paragon at NOAA/FSL is in conjunction with the ARPA-sponsored National Consortium for High-Performance Computing and as a part of the Boulder Front Range Consortium involving NCAR, the University of Colorado, and FSL.

Agency Ties	
ARPA	Partner
NSF	Partner
DOE	Partner
NASA	
NIH	
NSA	
NIST	
NOAA	
EPA	
ED	

FY 1993 Accomplishments	FY 1994 Major Milestones	FY 1995 Major Milestones	Milestone Changes from Prior IP
Running an extremely high resolution global ocean model, the Parallel Ocean Program (POP) on the DOE/LANL CM-5 as a result of a collaborative effort. R edesigned NOAA/GFDL SKYHI global atmospheric grid-point model, for parallel execution and ported to the DOE/LANL CM-5. Running a parallel, adiabatic version of the NOAA/NMC global spectral on the DOD/NRL CM-200 and CM-5. Most of the physical parameterizations for the model have been completed. The NMC ETA model restructured into parallel form has also been implemented on the DOD/NRL CM- 200. Installed a 208-node Intel Paragon at NOAA/FSL. FSL has developed a prototype parallel version of the Mesoscale Analysis and Prediction System (MAPS) high resolution weather model.	Complete preparations for a competitive procurement of a scalable HPCS (NOAA/GFDL). GFDL will also complete high resolution atmospheric modeling experiments using a parallel atmospheric grid model to test model sensitivity to grid resolution. Redesign and parallelize GFDL limited-area non-hydrostatic model to study the important climate problem of cloud-radiation interaction. Complete development of the physics for the global spectral model on the DOD/NRL CM-5 and make a comprehensive evaluation of its performance at very high resolution. Also complete testing of the restructured ETA model on the DOD/NRL CM-5 (NOAA/NMC). Design a restructured ETA model using FSL's "nearest neighbor" tool for implementation on the FSL Intel Paragon(NOAA/FSL,NOAA/NMC).	Complete procurement of a scalable HPCS to support major improvements in climate change and weather forecasting research . Begin scientific experiments using a parallel version of its SKYHI global atmospheric grid-point model. Begin scientific experiments using the parallel GFDL limited-area non-hydrostatic model to support research on radiation and clouds. Begin extensive testing of restructured, parallel version of the NMC regional/mesoscale ETA Optimum Interpolation (OI) code. Complete implementation of the restructured ETA model on the Intel Paragon. Begin extensive testing to evaluate its performance in terms of resolution and the potential impact for improved forecast guidance. Implement an advanced visualization laboratory to support high end applications .	

NOAA	Budget Account Code	Budget (BA, $ M)				HPCC Components by 1995 Pres. Req.				
		FY 93 Actual	FY 94 Pres.	FY 94 Est.	FY 95 Pres.	HPCS	NREN	ASTA	IITA	BRHR
Networking Connectivity		0.40	1.80	0.95	8.70		8.70			

Networking Connectivity is NOAA's program to greatly increase its ability to disseminate NOAA real-time and historical environmental data and information through the Internet to a broad range of users in the U.S. business community, government at all levels, research, education, and the general public.

Agency Ties	
ARPA	
NSF	
DOE	
NASA	
NIH	
NSA	
NIST	
NOAA	
EPA	
ED	

FY 1993 Accomplishments	FY 1994 Major Milestones	FY 1995 Major Milestones	Milestone Changes from Prior IP
Initial steps were taken to establish a NOAA Internet Network Information Center.	The number of NOAA computing and data archiving facilities at least minimally connected to the Internet will be increased by 10. One NOAA center of data will be connected to the Internet through a T3 connection. NOAA's Internet Network Information Center will be made operational.	At least 30 NOAA National Data Centers and Centers of Data will be connected to the Internet through a T1 or better connection. Five NOAA Centers of Data will be connected to the Internet through at least a 10 MBS communications link. A substantial amount of NOAA environmental data will be disseminated through the Internet to a wide range of users, using WAIS, Gopher, Mosaic servers and other advanced, user-supportive access tools.	

NOAA	Budget Account Code	Budget (BA, $ M)				HPCC Components by 1995 Pres. Req.				
		FY 93 Actual	FY 94 Pres.	FY 94 Est.	FY 95 Pres.	HPCS	NREN	ASTA	IITA	BRHR
Information Dissemination Pilots					0.50				0.50	

Information Dissemination Pilots is NOAA's program to begin pilot dissemination through the Internet of some of the vast amount of environmental data and information for which NOAA is responsible, and which is distributed geographically at NOAA data centers across the country, as part of the emerging National Information Infrastructure, for use by the private sector, academic researchers, educators, and the general public.

Agency Ties	
ARPA	
NSF	
DOE	
NASA	
NIH	
NSA	
NIST	
NOAA	
EPA	
ED	

FY 1993 Accomplishments	FY 1994 Major Milestones	FY 1995 Major Milestones	Milestone Changes from Prior IP
		Develop a data dissemination pilot that uses advanced data access tools on the Internet, and that makes selected heterogeneous NOAA environmental data which is distributed geographically at NOAA data centers, accessible to users in a more timely and complete way.	

Environmental Protection Agency

EPA's High Performance Computing and Communications (HPCC) program is focused on incorporating advances in computing and communications technology into key environmental assessment applications and transferring these advanced tools to key State, Federal, and industrial users. The integration of high performance technology enables complex multi-pollutant and cross- media pollutant assessments that are currently not feasible due to computational limitations. Early State pilot programs ensure the responsiveness of advanced tools to critical problems faced by State and local environmental groups. Research in distributed computing and massively parallel computer architectures provides for better access and management of large collections of environmental data and sharing of powerful computational resources needed to support advanced environmental problem solving. The current EPA program combines the efforts of personnel at EPA research laboratories and program offices, universities, technology centers, state environmental groups, and industrial partners. Some co-funding is provided by EPA air and information management programs. An external peer review of the EPA's HPCC program is scheduled for summer 1994.

FY 1995 funding for the EPA at the President's Request level is critical to achieving key FY 1995 milestones. Less than full funding will probably further delay the incorporation of parallel computing into the agency causing significant impacts on the effectiveness of the advanced multi-pollutant and cross-media assessment tools.

Program Activity	Budget Account Code	Partner/User Agencies	Budget (BA, $ M)				HPCC Components by 1995 Pres. Req.				
			FY 93 Actual	FY 94 Pres.	FY 94 Est.	FY 95 Pres.	HPCS	NREN	ASTA	IITA	BRHR
Environmental Modeling	CC9H1A, CC1A1E, CC2A1E, CA2H1A		3.33	5.15	5.03	6.45			6.45		
Computational Techniques	CC9H1A, CA2H1A	DOE	1.31	4.78	0.90	5.25			5.25		
State Network Connectivity	CC9H1A, CA2H1A, B4HX5A		0.21	0.70	0.70	0.70		0.70			
Education/Training	CC9H1A, B4HX5A		1.16	1.27	1.27	1.97					1.97
Public Data Access	CC9H1A	NASA, NOAA				0.30				0.30	
		Totals:	8.01	11.90	7.90	14.67		0.70	11.70	0.30	1.97

EPA	Budget Account Code	Budget (BA, $ M)				HPCC Components by 1995 Pres. Req.				
		FY 93 Actual	FY 94 Pres.	FY 94 Est.	FY 95 Pres.	HPCS	NREN	ASTA	IITA	BRHR
Environmental Modeling		5.33	5.15	5.03	6.45			6.45		

This program activity supports fundamental research at five universities and one state technology center on the systematic integration of advanced multi-pollutant environmental modeling components into a high performance distributed computing framework, addressing such issues as human-computer interface, distributed data management, software reuse and scalability, and system performance. EPA is integrating research results to develop a computational and decision support system for direct use by scientists, regulators, and industrial analysts involved in air pollution control. The initial focus of the program is on oxidant and acid deposition models, but will expand to encompass integration of more complex models such as aerosols, visibility and particulate. And the resultant decision support and data management framework will have general applicability for use in surface and ground water quality modeling and global tropospheric chemistry modeling. Advanced collaborative methods of analyzing and visualizing the multi-dimensional results from environmental assessment studies are being developed to provide a means of gaining greater insight into interactions of the science in the models and for better interpretation of results.

Agency Ties	
ARPA	
NSF	
DOE	
NASA	
NIH	
NSA	
NIST	
NOAA	
EPA	
ED	

FY 1993 Accomplishments	FY 1994 Major Milestones	FY 1995 Major Milestones	Milestone Changes from Prior IP
Two water quality and one air quality models ported to a supercomputer, and air model optimized for parallel performance. A new volume rendering algorithm enhanced the realism of three-dimensional model data visualization. Provided visualization tools with enhanced analysis capabilities to environmental scientists. Six cooperative research agreements were awarded for computational research on environmental modeling, decision support and visualization tools.	Develop concept and design of an air quality modeling decision support system that is easy to use and responsive to Federal, State, and industrial environmental problem solving needs. Demonstrate linked air and water quality models to enhance cross- media decision making. Develop interactive visualization techniques for analysis of multi-variate four-dimensional meteorological and air quality data.	Prototype modular, scalable air quality modeling decision support system in a distributed computing environment. Develop an interactive environment to facilitate the calibration, execution, and analysis of the Chesapeake Bay Watershed model for nitrogen eutrophication studies. Provide collaborative computing and visualization prototype for several air and water quality applications to enable distant researchers to co-examine data and effectively use remote visualization resources. Implement linked air & water quality models in a distributed computing environment to facilitate cross-media assessment of water quality.	

EPA	Budget Account Code	Budget (BA, $ M)				HPCC Components by 1995 Pres. Req.				
		FY 93 Actual	FY 94 Pres.	FY 94 Est.	FY 95 Pres.	HPCS	NREN	ASTA	IITA	BRHR
Computational Techniques		1.31	4.78	0.90	5.25			5.25		

This program activity performs parallel algorithm research and development at 5 universities and a DOE national laboratory. Research for EPA Grand Challenges such as air and water quality, molecular modeling, meteorological modeling, and stream and lake sedimentation is conducted on a variety of massively parallel architectures. The initial focus of the research is to determine which combinations of hardware, software and network capabilities will provide the required computational power to support the more complex multi-pollutant and cross-media modeling activities. Heterogeneous, distributed approaches are tested to ensure the most cost effective solutions. This research directly impacts the performance of EPA grand challenge applications and contributes to the shared software libraries and parallel testbed codes.

To fully integrate results of the initial efforts into the mainstream of EPA computational-based research activities, EPA plans to acquire a large scalable massively parallel computer. The scalability of the architectures allows an initial purchase of a relatively small machine with yearly additions to computing capacity.

Agency Ties	
ARPA	
NSF	
DOE	Partner
NASA	
NIH	
NSA	
NIST	
NOAA	
EPA	
ED	

FY 1993 Accomplishments	FY 1994 Major Milestones	FY 1995 Major Milestones	Milestone Changes from Prior IP
Regional Oxidant Model ported to a parallel SIMD architecture to serve as a testbed for parallel algorithm research.	Make available parallel codes for energy and force calculations for modeling chemical reactions and biological interaction of chemicals. Determine network bandwidth requirements for distributed environmental applications. Evaluate packetized video conferencing to support environmental applications.	Implement parallel codes for several gas-phase atmospheric chemistry solvers to evaluate performance of a variety of parallel architectures. Implement portable parallel meteorological testbed model to support advanced air quality models. Implement packetized video conferencing to key agency sites Acquire a scalable massively parallel (MP) computer to support HPCC research activities.	Acquisition of a scalable MP computer was delayed until FY 95 due to line item budget elimination from the FY-94 President's requested budget.

EPA	Budget Account Code	Budget (BA, $ M)				HPCC Components by 1995 Pres. Req.				
State Network Connectivity		FY 93 Actual	FY 94 Pres.	FY 94 Est.	FY 95 Pres.	HPCS	NREN	ASTA	IITA	BRHR
		0.21	0.70	0.70	0.70		0.70			

The main objectives of this program activity are to provide access to Internet for a critical mass of key environmental problem solving groups in State environmental communities to accelerate the wider use of more accurate and reliable environmental tools, and to provide capabilities for testing distributed computing approaches with complex cross-media environmental models. This activity supports network connections for five state and several federal pilot projects for evaluation of environmental assessment tools, grand challenge team research, and agency access to high performance computing.

This activity also provides high speed (45 million bits per second) interconnection through the Internet to the EPA's National Environmental Supercomputing Center in Bay City, Michigan with dedicated high speed connectivity to the EPA National Computing Center and HPCC research network in North Carolina. This interconnect establishes a basic foundation for the use of advanced environmental assessment tools throughout the agency and facilitates technology transfer to state environmental groups.

Agency Ties	
ARPA	
NSF	
DOE	
NASA	
NIH	
NSA	
NIST	
NOAA	
EPA	
ED	

FY 1993 Accomplishments	FY 1994 Major Milestones	FY 1995 Major Milestones	Milestone Changes from Prior IP
EPA established a 100 megabits/sec. (FDDI) research network in Research Triangle Park, NC and provided 1.5 megabits/sec. connectivity to the Chesapeake Bay Program Office in Maryland.	Provide 45 megabits/sec. interconnection through the Internet to the EPA's National Environmental Supercomputing Center. Provide a minimum of 56 kilobits/sec. connectivity to five state and one federal environmental groups to support HPCC pilot projects.	Demonstrate use of ATM technology for selected EPA applications. Maintain agency Internet and State connections, and upgrade selected state pilot project connections.	

EPA	Budget Account Code	Budget (BA, $ M)				HPCC Components by 1995 Pres. Req.				
		FY 93 Actual	FY 94 Pres.	FY 94 Est	FY 95 Pres.	HPCS	NREN	ASTA	IITA	BRHR
Education/Training		1.16	1.27	1.27	1.97					1.97

This program activity targets the dissemination of high performance computing and advanced environmental modeling capabilities to select groups from industry, state and federal users. The industrial group is represented by a consortium, the Consortium for Advanced Modeling of Regional Air Quality. The target groups of state users are selected from states with professionals involved in front-line environmental decision making and using models of regional and urban air quality. Pilot users gain first-hand experience about how high performance computing can alleviate many of their decision making constraints while at the same time they work with developers to ensure that system capabilities and interfaces address their needs. A high performance computing learning environment is created through workshops, involvement of professional trainers and curriculum developers, and establishment of a training "center" which functions remotely through Internet and NREN to support a distributed base of users of environmental models for decision making across the country.

Funding also supports EPA's EarthVision computational science educational program for high school students, graduate students, training fellowships, conferences and workshops.

Agency Ties	
ARPA	
NSF	
DOE	
NASA	
NIH	
NSA	
NIST	
NOAA	
EPA	
ED	

FY 1993 Accomplishments	FY 1994 Major Milestones	FY 1995 Major Milestones	Milestone Changes from Prior IP
Co-sponsorship of EarthVision, a computational science educational program for high school students and teachers. Cooperative research agreement awarded to explore curricular and training materials/methods related to transfer of advanced environmental assessment tools to a broad user community. Graduate students supported to provide cross discipline training in computational and environmental science	Continued co-sponsorship of EarthVision, a computational science educational program for high school students and teachers. Support environmental computational training fellowship Conduct workshops to assess training/educational needs for use of high performance environmental decision support tools	Prototype technology transfer methods for advanced air quality modeling systems and define concept for a distributed user/modeling support "center". Expand number of state pilot projects to obtain broader feedback on prototype decision support tools and training needs.	

EPA	Budget Account Code	Budget (BA, $ M)				HPCC Components by 1995 Pres. Req.				
		FY 93 Actual	FY 94 Pres.	FY 94 Est.	FY 95 Pres.	HPCS	NREN	ASTA	IITA	BRHR
Public Data Access					0.30				0.30	

EPA plans to expand public access to a variety of environmental databases, such as ecological measurements, air and water quality model predictions, population exposure to pollutants, etc., through development of intelligent user interfaces for interactive browsing, retrieval, analysis, and on-line multimedia tutorials for inexperienced users. EPA's current HPCC research addresses fundamental technology issues associated with the development of intelligent data access techniques and decision support tools for regulatory analysis and environmental education. The proposed IITA efforts expand the scope to satisfy the general public's earth science information needs.

Agency Ties	
ARPA	
NSF	
DOE	
NASA	Partner
NIH	
NSA	
NIST	
NOAA	Partner
EPA	
ED	

FY 1993 Accomplishments	FY 1994 Major Milestones	FY 1995 Major Milestones	Milestone Changes from Prior IP

ISBN 0-16-046307-6